INVERSE AND ILL-POSED PROBLEMS SERIES

Ill-Posed Internal Boundary Value Problems for the Biharmonic Equation

Also available in the Inverse and Ill-Posed Problems Series:

Investigation Methods for Inverse Problems
V.G. Romanov
Operator Theory. Nonclassical Problems
S.G. Pyatkov
Inverse Problems for Partial Differential Equations
Yu.Ya. Belov
Method of Spectral Mappings in the Inverse Problem Theory
V. Yurko
Theory of Linear Optimization
I.I. Eremin
Integral Geometry and Inverse Problems for Kinetic Equations
A.Kh. Amirov
Computer Modelling in Tomography and Ill-Posed Problems
M.M. Lavrent'ev, S.M. Zerkal and O.E. Trofimov
An Introduction to Identification Problems via Functional Analysis
A. Lorenzi
Coefficient Inverse Problems for Parabolic Type Equations and Their Application
P.G. Danilaev
Inverse Problems for Kinetic and Other Evolution Equations
Yu.E. Anikonov
Inverse Problems of Wave Processes
A.S. Blagoveshchenskii
Uniqueness Problems for Degenerating Equations and Nonclassical Problems
S.P. Shishatskii, A. Asanov and E.R. Atamanov
Uniqueness Questions in Reconstruction of Multidimensional Tomography-Type Projection Data
V.P. Golubyatnikov
Monte Carlo Method for Solving Inverse Problems of Radiation Transfer
V.S. Antyufeev
Introduction to the Theory of Inverse Problems
A.L. Bukhgeim
Identification Problems of Wave Phenomena - Theory and Numerics
S.I. Kabanikhin and A. Lorenzi
Inverse Problems of Electromagnetic Geophysical Fields
P.S. Martyshko
Composite Type Equations and Inverse Problems
A.I. Kozhanov
Inverse Problems of Vibrational Spectroscopy
A.G. Yagola, I.V. Kochikov, G.M. Kuramshina and Yu.A. Pentin
Elements of the Theory of Inverse Problems
A.M. Denisov
Volterra Equations and Inverse Problems
A.L. Bughgeim
Small Parameter Method in Multidimensional Inverse Problems
A.S. Barashkov
Regularization, Uniqueness and Existence of Volterra Equations of the First Kind
A. Asanov
Methods for Solution of Nonlinear Operator Equations
V.P. Tanana
Inverse and Ill-Posed Sources Problems
Yu.E. Anikonov, B.A. Bubnov and G.N. Erokhin
Methods for Solving Operator Equations
V.P. Tanana
Nonclassical and Inverse Problems for Pseudoparabolic Equations
A. Asanov and E.R. Atamanov
Formulas in Inverse and Ill-Posed Problems
Yu.E. Anikonov
Inverse Logarithmic Potential Problem
V.G. Cherednichenko
Multidimensional Inverse and Ill-Posed Problems for Differential Equations
Yu.E. Anikonov
Ill-Posed Problems with A Priori Information
V.V. Vasin and A.L. Ageev
Integral Geometry of Tensor Fields
V.A. Sharafutdinov
Inverse Problems for Maxwell's Equations
V.G. Romanov and S.I. Kabanikhin

INVERSE AND ILL-POSED PROBLEMS SERIES

Ill-Posed Internal Boundary Value Problems for the Biharmonic Equation

M.A. Atakhodzhaev

///VSP///

UTRECHT • BOSTON • KÖLN • TOKYO

2002

VSP
P.O. Box 346
3700 AH Zeist
The Netherlands

Tel: +31 30 692 5790
Fax: +31 30 693 2081
vsppub@compuserve.com
www.vsppub.com

First published in 2002

ISBN 90-6764-365-3

Printed in The Netherlands by Ridderprint bv, Ridderkerk.

Preface

By an internal boundary value problem we mean the problem of determining the solution of an equation if data are given on two manifolds. One manifold is the domain boundary, and the other manifold is situated inside the domain.

Here we study three essentially ill-posed internal boundary value problems for the biharmonic equation and the Cauchy problem for the abstract biharmonic equation.

This monograph is oriented to specialists in mathematical physics and to students and post-graduates majoring in this branch of mathematics.

The author is greatly indebted to his wife K. A. Atakhodzhaeva for her great help.

Contents

Introduction

Here by an internal boundary value problem we mean the problem of determining the solution of an equation if data are given on two manifolds. One of these manifolds is a part or the whole of the boundary of a domain. Another manifold is situated inside the domain.

Let D_j, $j = 1, 2$, be bounded domains on the plane, and let $D_1 \supset \overline{D_2}$, $\partial D_j = S_j$, $j = 1, 2$.

The first internal boundary value problem. We need to find a function $u(p)$, $p = (x, y)$, in D_1 which satisfies the conditions

$$\Delta^2 u(p) = 0, \qquad p \in D_1, \tag{1}$$

$$u(p) = \Phi_j(p), \qquad p \in S_j, \quad j = 1, 2. \tag{2}$$

Here Δ is the Laplace operator, and Φ_j are given functions.

The second internal boundary value problem. We need to find a function $u(p)$ in D_1 which satisfies conditions (1), (2) for $j = 2$, and the condition

$$\frac{\partial u(p)}{\partial n} = \Phi_1(p), \qquad p \in S_1, \tag{3}$$

where n is the external normal to S_1.

The third internal boundary value problem. Let S_1' be a part of the boundary S_1. We need to find a function $u(p)$ in D_1 which satisfies condition (1), condition (2) for $j = 2$, condition (2) for $j = 1$ with S_1' instead of S_1, and also the condition

$$\frac{\partial u(p)}{\partial n} = \Phi_3(p), \qquad p \in S_1'. \tag{4}$$

It is easy to see that if S_1 is a Lyapunov contour, then the first internal boundary value problem (IBVP 1) and the second internal boundary value problem (IBVP 2) are equivalent to a system of integral equations of third kind, and the third internal boundary value problem (IBVP 3) is equivalent to a system of Fredholm integral equations of first kind. Therefore, all these problems are essentially ill-posed (Tikhonov and Arsenin, 1977).

In Chapters 1–3 of the present monograph, we study the above-mentioned three problems (both qualitatively and quantitatively). Chapter 4 deals with some abstract variants of these three problems and the Cauchy problem. The last chapter presents the m-dimensional case. In this monograph, the author introduces some new notions, for example, the notion of complete solvability.

Chapter 1.

The first internal boundary value problem

Let S_1 be a Lyapunov contour. Then the solution of equation (1) may be sought as the sum of a double layer potential and a logarithmic potential over the domain D_1

$$u(p) = \int_{S_1} \mu_1(q) \frac{\partial}{\partial n_q} \ln |p-q| \, \mathrm{d}_q S_1 + \int_{D_1} V(Q) \ln |p-Q| \, \mathrm{d}Q, \qquad (1.0.1)$$

where

$$V(Q) = \int_{S_1} \mu_2(q) \frac{\partial}{\partial n_q} \ln |q-Q| \, \mathrm{d}_q S_1.$$

Then, taking into account condition (2), we obtain from (1.0.1) the following system of Fredholm integral equations of third kind

$$\begin{aligned} -\pi\mu_1(p) + \int_{S_1} \mu_1(q) \frac{\partial}{\partial n_q} \ln |p-q| \, \mathrm{d}_q S_1 & \\ + \int_{S_1} \mu_2(q) K(p,q) \, \mathrm{d}_q S_1 &= \Phi_1(p), \quad p \in S_1, \\ \int_{S_1} \mu_1(q) \frac{\partial}{\partial n_q} \ln |p-q| \, \mathrm{d}_q S_1 & \\ + \int_{S_1} \mu_2(q) K(p,q) \, \mathrm{d}_q S_1 &= \Phi_2(p), \quad p \in S_2, \end{aligned} \qquad (1.0.2)$$

where

$$K(p,q) = \int_{D_1} \ln |p-Q| \frac{\partial}{\partial n_q} \ln |q-Q| \, \mathrm{d}Q.$$

Conversely, if $\mu_j(p) \in C(S_1)$, $j = 1, 2$, then, substituting them into (1.0.1), we find the solution of (1), (2).

Theorem 1.0.1. *If S_2 cannot be represented as a finite sum of analytic curves, then the solution of the problem* (1), (2) *is unique.*

Proof. Since the solution $u(x, y)$ of the biharmonic equation is an analytic function of two real variables (x, y), therefore, it can be continued analytically into a certain complex neighborhood. In the theory of analytic functions of several complex arguments, the following statement holds:

Let $f(x, y)$ be an analytic function of two real variables in a domain D_1, the function f admitting an analytic continuation into some complex neighborhood of the domain D_1. Suppose that $f(x, y) \not\equiv 0$ in D_3 $(D_3 \subset D_1)$ and A is the set of all those points in D_3 at which $f(x, y) = 0$ $((x, y) \in A)$. Then $A = \bigcup_{k=1}^{n} \Gamma_k$ where Γ_k are analytic curves.

Using this statement we see that since $S_2 \neq A$, therefore, condition $\Phi_2 = 0$ on S_2 implies $u(x, y) = 0$ in D_1. □

Remark 1.0.1. The above theorem does not require the analytic function to vanish on S_1.

Suppose that D is a star-like domain relative to the origin and its boundary S is a Lyapunov contour. We suppose also that our domain partially intersects with a disk D_R with the boundary S_R. We will denote by S^e, S_R^e and S^i, S_R^i, respectively, the external and internal parts of S and S_R. Let

$$D_1 = D \cup D_R, \qquad D_2 = D \cap D_R, \qquad \partial D_j = S_j, \quad j = 1, 2,$$
$$S_1 = S^e \cup S_R^e, \qquad S_2 = S^i \cup S_R^i.$$

In particular, if $D \subset D_R$, then $D_1 = D_R$ and $D_2 = D$; if $D \supset D_R$, then $D_1 = D$ and $D_2 = D_R$.

Theorem 1.0.2. *A solution of the problem* (1), (2) *is also unique if $D_1 = D \cup D_R$ and $D_2 = D \cap D_R$.*

Proof. As is known, the general solution of equation (1) in a star-like domain D_1 is as follows

$$u(r, \varphi) = u_1(r, \varphi) + (R^2 - r^2)\, u_2(r, \varphi), \tag{1.0.3}$$

where u_j, $j = 1, 2$, are harmonic functions.

Now assume that $u^{(1)}$ and $u^{(2)}$ are two functions satisfying conditions (1) and (2). Then their difference $u = u^{(1)} - u^{(2)}$ is equal to zero on S_j, $j = 1, 2$, and $u_1 = 0$ on S_R, $u_2 = 0$ on S. Hence it follows that $u_1 = 0$ in D_R and $u_2 = 0$ in D, since the internal Dirichlet problem has a unique solution. Thus, $u_j = 0$, $j = 1, 2$, in D_1, which follows from the uniqueness of harmonic continuation. Hence, taking into account (1.0.3), we have $u = 0$ in D_1. Therefore, $u^{(1)} = u^{(2)}$. Note that for arbitrary domains D_1 and D_2 a solution of the problem (1), (2) may be not unique. □

1.1. AN EXAMPLE OF NONUNIQUENESS

We suppose that D_1 is a star-like domain and consider the function

$$V(r, \varphi) = r^2 \int_{S_1} f_2 \frac{\partial G_1}{\partial n} \, \mathrm{d}S_1 - \int_{S_1} R_1^2 f_2 \frac{\partial G_1}{\partial n} \, \mathrm{d}S_1, \tag{1.1.1}$$

where $r = R_1(\varphi)$ is the equation for S_1. It is easy to see that the function V satisfies equation (1) and is equal to zero on S_1 for each continuous function f_2. This is so because G_1 is the Green function of the Dirichlet problem for the Laplace equation. Now we consider the expansion of the function $V(r, \varphi)$ in the Taylor series in a neighborhood of the origin. Let us write the result in the form

$$\begin{aligned} V(r, \varphi) = A_0 + (A_1 \cos\varphi + B_1 \sin\varphi) r + C_0 r^2 \\ + (A_2 \cos 2\varphi + B_2 \sin 2\varphi) r^2 + V_1(r, \varphi) r^3. \end{aligned} \tag{1.1.2}$$

Here V_1 is some analytic function. The coefficients A_0, A_1, A_2, B_1, B_2, and C_0 in (1.1.2) are linear functionals of the function $f_2(\psi)$:

$$\begin{aligned} A_0 &= \int_0^{2\pi} \psi_0(\psi) f_2(\psi) \, \mathrm{d}\psi, & A_1 &= \int_0^{2\pi} \psi_1(\psi) f_2(\psi) \, \mathrm{d}\psi, \\ A_2 &= \int_0^{2\pi} \psi_3(\psi) f_2(\psi) \, \mathrm{d}\psi, & C_0 &= \int_0^{2\pi} \psi_5(\psi) f_2(\psi) \, \mathrm{d}\psi, \\ B_1 &= \int_0^{2\pi} \psi_2(\psi) f_2(\psi) \, \mathrm{d}\psi, & B_2 &= \int_0^{2\pi} \psi_4(\psi) f_2(\psi) \, \mathrm{d}\psi. \end{aligned} \tag{1.1.3}$$

The functions $\psi_j(\psi)$ $(j = \overline{0, 5})$ depend on the domain D_1.

Suppose that for some domain D_1, the system of functions $\{\psi_j\}_0^5$ is linearly independent. Evidently, in this case, for each set of numbers

$$A_0^0, \quad A_1^0, \quad A_2^0, \quad B_1^0, \quad B_2^0, \quad C_0^0,$$

there exists a function f_2 such that

$$A_0 = A_0^0, \quad A_1 = A_1^0, \quad A_2 = A_2^0, \quad B_1 = B_1^0, \quad B_2 = B_2^0, \quad C_0 = C_0^0.$$

In particular, there exists a function f_2 such that formula (1.1.2) becomes

$$V(r,\varphi) = A_0 + C_0 r^2 + V_1(r,\varphi) r^3. \tag{1.1.4}$$

We suppose now that r_1 is some sufficiently small number. Let f_2 be a function such that the coefficients in (1.1.4) are as follows:

$$A_0 = r_1^2, \qquad C_0 = -2.$$

The function $V_1(r,\varphi)$ in (1.1.4) is defined by the function f_2 and may be considered as a bounded function. Therefore, for r_1 sufficiently small, we have

$$\begin{aligned} &V(0,\varphi) = r_1^2 > 0, \\ &V(r_1,\varphi) = -r_1^2 + V_1(r_1,\varphi) r_1^3 < 0, \\ &V_r'(r,\varphi) = -4r + 3r^2 \Big[V_1(r,\varphi) + \frac{r}{3} V_{1r}'(r,\varphi) \Big] < 0, \qquad r < r_1. \end{aligned}$$

Therefore, in the disk $r < r_1$ there exists a curve S_2 such that the function $V(r,\varphi)$ vanishes on S_2. Let $r = R_2(\varphi)$ be the equation of S_2. We have $R_2(\varphi) < r_1$ and

$$V(R_2(\varphi), \varphi) = 0.$$

Thus, we have shown that if the domain D_1 is such that the functions $\{\psi_j\}_0^5$ are linearly independent, then there exists a closed curve S_2 such that the solution of (1), (2) is not unique. Note that, if D_1 is a disk, then

$$\psi_0(\psi) = C_1 \psi_5(\psi),$$

where C_1 is some constant.

Now let us show that there exist domains which are close to a disk and whose systems of functions $\{\psi_j\}_0^5$ are linearly independent. Suppose that S_1 has the form

$$r = R_1(\varphi) = 1 - \varepsilon \cos 3\varphi, \qquad 0 \le \varphi \le 2\pi, \tag{1.1.5}$$

where ε is a fixed sufficiently small number.

Let $f(z)$ denote the analytic function of the complex variable $z = re^{i\varphi}$ which maps D_1 onto the unit disk. Then the function (1.1.1) can be written as follows (Vladimirov, 1984)

$$V(r,\varphi) = \operatorname{Re}\frac{1}{2\pi i}\int_{S_1}[r^2 - R_1^2(\psi)]f_2(\zeta)J(\zeta,z)\,d\zeta, \tag{1.1.6}$$

where

$$J(\zeta,z) = \frac{f(\zeta)+f(z)}{f(\zeta)-f(z)}\cdot\frac{f'(\zeta)}{f(\zeta)},$$
$$z = re^{i\varphi}, \quad \zeta = (1-\varepsilon\cos 3\psi)e^{i\psi},$$
$$d\zeta = \left[i + \varepsilon(3\sin 3\psi - i\cos 3\psi)\right]e^{i\psi}\,d\psi.$$

Using the approximate formula of conformal mapping (Lavrent'ev and Shabat, 1966, pp. 371–379), we can write

$$f(z) = z + \varepsilon z^4 + O(\varepsilon^2), \qquad f'(\zeta) = 1 + 4\varepsilon\zeta^3 + O(\varepsilon^2). \tag{1.1.7}$$

Below we calculate to within $O(\varepsilon^2)$ and $O(r^3)$. As a result we obtain

$$J(\zeta,z) \approx \frac{\zeta + z + \varepsilon(\zeta^4 + z^4)}{\zeta - z + \varepsilon(\zeta^4 - z^4)}\cdot\frac{1+4\varepsilon\zeta^3}{\zeta(1+\varepsilon\zeta^3)}, \tag{1.1.8}$$

$$\frac{1}{\zeta - z + \varepsilon(\zeta^4 - z^4)} = \frac{1}{\zeta - z}\cdot\frac{1}{1+\varepsilon\dfrac{\zeta^4 - z^4}{\zeta - z}} \approx \frac{1}{\zeta - z}\left[1 - \varepsilon\frac{\zeta^4 - z^4}{\zeta - z}\right]$$
$$= \frac{1}{\zeta - z}\left[1 - \varepsilon\frac{(\zeta^2 - z^2)(\zeta^2 + z^2)}{\zeta - z}\right] = \frac{1}{\zeta - z}[1 - \varepsilon(\zeta + z)(\zeta^2 + z^2)],$$

$$\frac{1}{\zeta[1+\varepsilon\zeta^3]} \approx \frac{1}{\zeta}[1 - \varepsilon\zeta^3],$$

$$\left[\zeta + z + \varepsilon(\zeta^4 + z^4)\right](1 + 4\varepsilon\zeta^3) \approx (\zeta + z)\left[1 + 4\varepsilon\zeta^3 + \frac{\zeta^4 + z^4}{\zeta + z}\varepsilon\right],$$

$$\frac{\zeta^4 + z^4}{\zeta + z} = \frac{\zeta^4 - z^4 + 2z^4}{\zeta + z} = \zeta^3 - \zeta^2 z + \zeta z^2 - z^3 + \frac{2z^4}{\zeta + z},$$

$$J(\zeta,z) \approx \frac{\zeta + z}{\zeta(\zeta - z)}\left[1 + 4\varepsilon\zeta^3 + \varepsilon\zeta^3 - \varepsilon\zeta^2 z + \varepsilon\zeta z^2 - \varepsilon z^3 + \frac{2\varepsilon z^4}{\zeta + z}\right]$$
$$\times\left[1 - \varepsilon(\zeta + z)(\zeta^2 + z^2)\right](1 - \varepsilon\zeta^3)$$

$$\approx \frac{\zeta+z}{\zeta(\zeta-z)}\Big[1+5\varepsilon\zeta^3-\varepsilon\zeta^2 z+\varepsilon\zeta z^2-\varepsilon z^3+\frac{2\varepsilon z^4}{\zeta+z}-\varepsilon\zeta^3$$

$$-\varepsilon z\zeta^2-\varepsilon\zeta z^2-\varepsilon z^3-\varepsilon\zeta^3\Big]$$

$$=\frac{\zeta+z}{\zeta(\zeta-z)}\Big[1+\varepsilon\Big(3\zeta^3-2z\zeta^2-2z^3+\frac{2z^4}{\zeta+z}\Big)\Big],$$

$$\zeta^2\approx(1-2\varepsilon\cos 3\psi)\mathrm{e}^{2\mathrm{i}\psi},\qquad \zeta^3\approx(1-3\varepsilon\cos 3\psi)\mathrm{e}^{3\mathrm{i}\psi},$$

$$\frac{1}{\zeta}=\frac{1}{(1-\varepsilon\cos 3\psi)e^{i\psi}}\approx\frac{1+\varepsilon\cos 3\psi}{e^{i\psi}},$$

$$\zeta+z=(1-\varepsilon\cos 3\psi)e^{i\psi}+r\mathrm{e}^{\mathrm{i}\varphi}=(\mathrm{e}^{\mathrm{i}\psi}+r\mathrm{e}^{\mathrm{i}\varphi})\Big(1-\frac{\varepsilon\cos 3\psi\cdot\mathrm{e}^{\mathrm{i}\psi}}{\mathrm{e}^{\mathrm{i}\psi}+r\mathrm{e}^{\mathrm{i}\varphi}}\Big),$$

$$\frac{1}{\zeta-z}=\frac{1}{(1-\varepsilon\cos 3\psi)\mathrm{e}^{\mathrm{i}\psi}-r\mathrm{e}^{\mathrm{i}\varphi}}=\frac{1}{(\mathrm{e}^{\mathrm{i}\psi}-r\mathrm{e}^{\mathrm{i}\varphi})\Big[1-\dfrac{\varepsilon\cos 3\psi\cdot\mathrm{e}^{\mathrm{i}\psi}}{\mathrm{e}^{\mathrm{i}\psi}-r\mathrm{e}^{\mathrm{i}\varphi}}\Big]}$$

$$\approx\frac{1}{\mathrm{e}^{\mathrm{i}\psi}-r\mathrm{e}^{\mathrm{i}\varphi}}\Big[1+\frac{\varepsilon\cos 3\psi\cdot\mathrm{e}^{\mathrm{i}\psi}}{\mathrm{e}^{\mathrm{i}\psi}-r\mathrm{e}^{\mathrm{i}\varphi}}\Big],$$

$$\frac{\zeta+z}{\zeta(\zeta-z)}\,\mathrm{d}\zeta\approx(1+\varepsilon\cos 3\psi)(\mathrm{e}^{\mathrm{i}\psi}+r\mathrm{e}^{\mathrm{i}\varphi})\Big(1-\frac{\varepsilon\cos 3\psi\cdot\mathrm{e}^{\mathrm{i}\psi}}{\mathrm{e}^{\mathrm{i}\psi}+r\mathrm{e}^{\mathrm{i}\varphi}}\Big)$$

$$\times\frac{1}{\mathrm{e}^{\mathrm{i}\psi}-r\mathrm{e}^{\mathrm{i}\varphi}}\Big[1+\frac{\varepsilon\cos 3\psi\cdot\mathrm{e}^{\mathrm{i}\psi}}{\mathrm{e}^{\mathrm{i}\psi}-r\mathrm{e}^{\mathrm{i}\varphi}}\Big]\big[\mathrm{i}+\varepsilon(3\sin 3\psi-\mathrm{i}\cos 3\psi)\big]\,\mathrm{d}\psi$$

$$\approx\frac{\mathrm{e}^{\mathrm{i}\psi}+r\mathrm{e}^{\mathrm{i}\varphi}}{\mathrm{e}^{\mathrm{i}\psi}-r\mathrm{e}^{\mathrm{i}\varphi}}\,\mathrm{d}\psi\,\Big[\mathrm{i}+\mathrm{i}\varepsilon\cos 3\psi-\frac{\mathrm{i}\varepsilon\cos 3\psi\cdot\mathrm{e}^{\mathrm{i}\psi}}{\mathrm{e}^{\mathrm{i}\psi}+r\mathrm{e}^{\mathrm{i}\varphi}}+\frac{\mathrm{i}\varepsilon\cos 3\psi\cdot\mathrm{e}^{\mathrm{i}\psi}}{\mathrm{e}^{\mathrm{i}\psi}-r\mathrm{e}^{\mathrm{i}\varphi}}$$

$$+\varepsilon(3\sin 3\psi-\mathrm{i}\cos 3\psi)\Big]$$

$$=\frac{\mathrm{e}^{\mathrm{i}\psi}+r\mathrm{e}^{\mathrm{i}\varphi}}{\mathrm{e}^{\mathrm{i}\psi}-r\mathrm{e}^{\mathrm{i}\varphi}}\,\mathrm{d}\psi\,\Big[\mathrm{i}+3\varepsilon\sin 3\psi+\mathrm{i}\varepsilon\cos 3\psi\cdot\mathrm{e}^{\mathrm{i}\psi}\Big(\frac{1}{\mathrm{e}^{\mathrm{i}\psi}-r\mathrm{e}^{\mathrm{i}\varphi}}-\frac{1}{\mathrm{e}^{\mathrm{i}\psi}+r\mathrm{e}^{\mathrm{i}\varphi}}\Big)\Big]$$

$$=\mathrm{i}\,\frac{\mathrm{e}^{\mathrm{i}\psi}+r\mathrm{e}^{\mathrm{i}\varphi}}{\mathrm{e}^{\mathrm{i}\psi}-r\mathrm{e}^{\mathrm{i}\varphi}}\,\mathrm{d}\psi+3\varepsilon\sin 3\psi\,\frac{\mathrm{e}^{\mathrm{i}\psi}+r\mathrm{e}^{\mathrm{i}\varphi}}{\mathrm{e}^{\mathrm{i}\psi}-r\mathrm{e}^{\mathrm{i}\varphi}}\,\mathrm{d}\psi+\frac{2\mathrm{i}\varepsilon r\cos 3\psi\cdot\mathrm{e}^{\mathrm{i}(\varphi+\psi)}}{(\mathrm{e}^{\mathrm{i}\psi}-r\mathrm{e}^{\mathrm{i}\varphi})^2}\,\mathrm{d}\psi,$$

$$1+\varepsilon\Big(3\zeta^3-2z\zeta^2-2z^3+\frac{2z^4}{\zeta+z}\Big)\approx 1+\varepsilon(3\zeta^3-2z\zeta^2)$$

$$\approx 1+\varepsilon\big[3(1-3\varepsilon\cos 3\psi)\mathrm{e}^{3\mathrm{i}\psi}-2r\mathrm{e}^{\mathrm{i}\varphi}(1-2\varepsilon\cos 3\psi)\mathrm{e}^{2\mathrm{i}\psi}\big]$$

$$\approx 1+\varepsilon\big[3\mathrm{e}^{3\mathrm{i}\psi}-2r\mathrm{e}^{\mathrm{i}(\varphi+2\psi)}\big],$$

$$J(\zeta,z)\,\mathrm{d}\zeta \approx \left[\mathrm{i}\frac{\mathrm{e}^{\mathrm{i}\psi}+r\mathrm{e}^{\mathrm{i}\varphi}}{\mathrm{e}^{\mathrm{i}\psi}-r\mathrm{e}^{\mathrm{i}\varphi}} + 3\varepsilon\sin 3\psi\,\frac{\mathrm{e}^{\mathrm{i}\psi}+r\mathrm{e}^{\mathrm{i}\varphi}}{\mathrm{e}^{\mathrm{i}\psi}-r\mathrm{e}^{\mathrm{i}\varphi}} + \frac{2\mathrm{i}\varepsilon r\cos 3\psi\cdot\mathrm{e}^{\mathrm{i}(\varphi+\psi)}}{(\mathrm{e}^{\mathrm{i}\psi}-r\mathrm{e}^{\mathrm{i}\varphi})^2}\right]$$
$$\times\left[1+\varepsilon(3\mathrm{e}^{3\mathrm{i}\psi}-2r\mathrm{e}^{\mathrm{i}(\varphi+2\psi)})\right]\mathrm{d}\psi$$
$$\approx \mathrm{i}\frac{\mathrm{e}^{\mathrm{i}\psi}+r\mathrm{e}^{\mathrm{i}\varphi}}{\mathrm{e}^{\mathrm{i}\psi}-r\mathrm{e}^{\mathrm{i}\varphi}}\,\mathrm{d}\psi + 3\varepsilon\sin 3\psi\,\frac{\mathrm{e}^{\mathrm{i}\psi}+r\mathrm{e}^{\mathrm{i}\varphi}}{\mathrm{e}^{\mathrm{i}\psi}-r\mathrm{e}^{\mathrm{i}\varphi}}\,\mathrm{d}\psi + \frac{2\mathrm{i}\varepsilon r\cos 3\psi\cdot\mathrm{e}^{\mathrm{i}(\varphi+\psi)}}{(\mathrm{e}^{\mathrm{i}\psi}-r\mathrm{e}^{\mathrm{i}\varphi})^2}\,\mathrm{d}\psi$$
$$+\,\mathrm{i}\varepsilon\frac{\mathrm{e}^{\mathrm{i}\psi}+r\mathrm{e}^{\mathrm{i}\varphi}}{\mathrm{e}^{\mathrm{i}\psi}-r\mathrm{e}^{\mathrm{i}\varphi}}\left[3\mathrm{e}^{3\mathrm{i}\psi}-2r\mathrm{e}^{\mathrm{i}(\varphi+2\psi)}\right]\mathrm{d}\psi,$$
$$(\mathrm{e}^{\mathrm{i}\psi}+r\mathrm{e}^{\mathrm{i}\varphi})(3\mathrm{e}^{3\mathrm{i}\psi}-2r\mathrm{e}^{\mathrm{i}(\varphi+2\psi)})$$
$$=3\mathrm{e}^{4\mathrm{i}\psi}+3r\mathrm{e}^{\mathrm{i}(\varphi+3\psi)}-2r\mathrm{e}^{\mathrm{i}(\varphi+3\psi)}-2r^2\mathrm{e}^{2\mathrm{i}(\varphi+\psi)}.$$

Therefore,

$$J(\zeta,z)\,\mathrm{d}\zeta \approx \mathrm{i}\frac{\mathrm{e}^{\mathrm{i}\psi}+r\mathrm{e}^{\mathrm{i}\varphi}}{\mathrm{e}^{\mathrm{i}\psi}-r\mathrm{e}^{\mathrm{i}\varphi}}\,\mathrm{d}\psi + 3\varepsilon\sin 3\psi\,\frac{\mathrm{e}^{\mathrm{i}\psi}+r\mathrm{e}^{\mathrm{i}\varphi}}{\mathrm{e}^{\mathrm{i}\psi}-r\mathrm{e}^{\mathrm{i}\varphi}}\,\mathrm{d}\psi$$
$$+\frac{2\mathrm{i}\varepsilon r\cos 3\psi\cdot\mathrm{e}^{\mathrm{i}(\varphi+\psi)}}{(\mathrm{e}^{\mathrm{i}\psi}-r\mathrm{e}^{\mathrm{i}\varphi})^2}\,\mathrm{d}\psi$$
$$+\frac{\mathrm{i}\varepsilon\,\mathrm{d}\psi}{\mathrm{e}^{\mathrm{i}\psi}-r\mathrm{e}^{\mathrm{i}\varphi}}\left(3\mathrm{e}^{4\mathrm{i}\psi}+r\mathrm{e}^{\mathrm{i}(\varphi+3\psi)}-2r^2\mathrm{e}^{2\mathrm{i}(\varphi+\psi)}\right).$$

Further, we have

$$r^2-R_1^2(\psi)\approx r^2-1+2\varepsilon\cos 3\psi,$$
$$[r^2-R_1^2(\psi)]J(\zeta,z)\,\mathrm{d}\zeta \approx (r^2-1+2\varepsilon\cos 3\psi)$$
$$\times\left\{\mathrm{i}\frac{\mathrm{e}^{\mathrm{i}\psi}+r\mathrm{e}^{\mathrm{i}\varphi}}{\mathrm{e}^{\mathrm{i}\psi}-r\mathrm{e}^{\mathrm{i}\varphi}}\,\mathrm{d}\psi + 3\varepsilon\sin 3\psi\,\frac{\mathrm{e}^{\mathrm{i}\psi}+r\mathrm{e}^{\mathrm{i}\varphi}}{\mathrm{e}^{\mathrm{i}\psi}-r\mathrm{e}^{\mathrm{i}\varphi}}\,\mathrm{d}\psi + \frac{2\mathrm{i}\varepsilon r\cos 3\psi\cdot\mathrm{e}^{\mathrm{i}(\varphi+\psi)}}{(\mathrm{e}^{\mathrm{i}\psi}-r\mathrm{e}^{\mathrm{i}\varphi})^2}\,\mathrm{d}\psi\right.$$
$$\left.+\frac{\mathrm{i}\varepsilon\,\mathrm{d}\psi}{\mathrm{e}^{\mathrm{i}\psi}-r\mathrm{e}^{\mathrm{i}\varphi}}\left(3\mathrm{e}^{4\mathrm{i}\psi}+r\mathrm{e}^{\mathrm{i}(\varphi+3\psi)}-2r^2\mathrm{e}^{2\mathrm{i}(\varphi+\psi)}\right)\right\}$$
$$\approx \mathrm{i}(r^2-1)\frac{\mathrm{e}^{\mathrm{i}\psi}+r\mathrm{e}^{\mathrm{i}\varphi}}{\mathrm{e}^{\mathrm{i}\psi}-r\mathrm{e}^{\mathrm{i}\varphi}}\,\mathrm{d}\psi + 2\mathrm{i}\varepsilon\cos 3\psi\,\frac{\mathrm{e}^{\mathrm{i}\psi}+r\mathrm{e}^{\mathrm{i}\varphi}}{\mathrm{e}^{\mathrm{i}\psi}-r\mathrm{e}^{\mathrm{i}\varphi}}\,\mathrm{d}\psi$$
$$+3(r^2-1)\varepsilon\sin 3\psi\,\frac{\mathrm{e}^{\mathrm{i}\psi}+r\mathrm{e}^{\mathrm{i}\varphi}}{\mathrm{e}^{\mathrm{i}\psi}-r\mathrm{e}^{\mathrm{i}\varphi}}\,\mathrm{d}\psi + (r^2-1)\frac{2\mathrm{i}\varepsilon r\cos 3\psi\cdot\mathrm{e}^{\mathrm{i}(\varphi+\psi)}}{(\mathrm{e}^{\mathrm{i}\psi}-r\mathrm{e}^{\mathrm{i}\varphi})^2}\,\mathrm{d}\psi$$
$$+\frac{\mathrm{i}\varepsilon(r^2-1)\,\mathrm{d}\psi}{\mathrm{e}^{\mathrm{i}\psi}-r\mathrm{e}^{\mathrm{i}\varphi}}\left(3\mathrm{e}^{4\mathrm{i}\psi}+r\mathrm{e}^{\mathrm{i}(\varphi+3\psi)}-2r^2\mathrm{e}^{2\mathrm{i}(\varphi+\psi)}\right),$$

$$\frac{1}{e^{i\psi} - re^{i\varphi}} = \frac{e^{-i\psi}}{1 - re^{i(\varphi-\psi)}} \approx e^{-i\psi}\big(1 + re^{i(\varphi-\psi)} + r^2 e^{2i(\varphi-\psi)}\big)$$

$$= e^{-i\psi} + re^{i(\varphi-2\psi)} + r^2 e^{i(2\varphi-3\psi)},$$

$$\frac{1}{(e^{i\psi} - re^{i\varphi})^2} \approx e^{-2i\psi} + r^2 e^{2i(\varphi-2\psi)} + 2re^{i(\varphi-3\psi)} + 2r^2 e^{i(2\varphi-4\psi)}$$

$$= e^{-2i\psi} + 2re^{i(\varphi-3\psi)} + 3r^2 e^{i(2\varphi-4\psi)},$$

$$\frac{e^{i\psi} + re^{i\varphi}}{e^{i\psi} - re^{i\varphi}} \approx (e^{i\psi} + re^{i\varphi})\big(e^{-i\psi} + re^{i(\varphi-2\psi)} + r^2 e^{i(2\varphi-3\psi)}\big)$$

$$\approx 1 + re^{i(\varphi-\psi)} + re^{i(\varphi-\psi)} + r^2 e^{i(2\varphi-2\psi)} + r^2 e^{i(2\varphi-2\psi)}$$

$$= 1 + 2re^{i(\varphi-\psi)} + 2r^2 e^{2i(\varphi-\psi)},$$

$$\frac{r^2-1}{e^{i\psi} - re^{i\varphi}} \approx (r^2-1)\big(e^{-i\psi} + re^{i(\varphi-2\psi)} + r^2 e^{i(2\varphi-3\psi)}\big)$$

$$\approx r^2 e^{-i\psi} - e^{-i\psi} - re^{i(\varphi-2\psi)} - r^2 e^{i(2\varphi-3\psi)},$$

$$(r^2-1)\frac{e^{i\psi} + re^{i\varphi}}{e^{i\psi} - re^{i\varphi}} \approx (r^2-1)\big(1 + 2re^{i(\varphi-\psi)} + 2r^2 e^{2i(\varphi-\psi)}\big)$$

$$\approx r^2 - 1 - 2re^{i(\varphi-\psi)} - 2r^2 e^{2i(\varphi-\psi)},$$

$$\frac{r^2-1}{(e^{i\psi} - re^{i\varphi})^2} \approx (r^2-1)\big(e^{-2i\psi} + 2re^{i(\varphi-3\psi)} + 3r^2 e^{2i(\varphi-2\psi)}\big)$$

$$\approx r^2 e^{-2i\psi} - e^{-2i\psi} - 2re^{i(\varphi-3\psi)} - 3r^2 e^{2i(\varphi-2\psi)},$$

$$\frac{r^2-1}{e^{i\psi} - re^{i\varphi}}\big(3e^{4i\psi} + re^{i(\varphi+3\psi)} - 2r^2 e^{2i(\varphi+\psi)}\big)$$

$$\approx \big(r^2 e^{-i\psi} - e^{-i\psi} - re^{i(\varphi-2\psi)} - r^2 e^{i(2\varphi-3\psi)}\big)$$

$$\times \big(3e^{4i\psi} + re^{i(\varphi+3\psi)} - 2r^2 e^{2i(\varphi+\psi)}\big)$$

$$\approx 3r^2 e^{3i\psi} - 3e^{3i\psi} - re^{i(\varphi+2\psi)} + 2r^2 e^{i(2\varphi+\psi)}$$

$$- 3re^{i(\varphi+2\psi)} - r^2 e^{i(2\varphi+\psi)} - 3r^2 e^{i(2\varphi+\psi)}$$

$$= -3e^{3i\psi} - 4re^{i(\varphi+2\psi)} + 3r^2 e^{3i\psi} - 2r^2 e^{i(2\varphi+\psi)}.$$

Thus,

$$[r^2 - R_1^2(\psi)]J(\zeta, z)\,d\zeta \approx d\psi\Big\{ir^2 - i - 2ire^{i(\varphi-\psi)} - 2ir^2 e^{2i(\varphi-\psi)}$$

$$+ 2i\varepsilon\cos 3\psi + 4i\varepsilon r\cos 3\psi\cdot e^{i(\varphi-\psi)} + 4i\varepsilon r^2\cos 3\psi\cdot e^{2i(\varphi-\psi)} + 3\varepsilon r^2\sin 3\psi$$
$$- 3\varepsilon\sin 3\psi - 6\varepsilon r\sin 3\psi\cdot e^{i(\varphi-\psi)} - 6\varepsilon r^2\sin 3\psi\cdot e^{2i(\varphi-\psi)}$$
$$- 2i\varepsilon r\cos 3\psi\cdot e^{i(\varphi-\psi)} - 4i\varepsilon r^2\cos 3\psi\cdot e^{i(2\varphi-2\psi)} - 3i\varepsilon e^{3i\psi} - 4i\varepsilon r e^{i(\varphi+2\psi)}$$
$$+ 3i\varepsilon r^2 e^{3i\psi} - 2i\varepsilon r^2 e^{i(2\varphi+\psi)}\Big\},$$

$$V(r,\varphi) \approx \operatorname{Re}\frac{1}{2\pi i}\int_0^{2\pi} f_2(\psi)\,d\psi\,\Big\{ir^2 - i - 2ire^{i(\varphi-\psi)} - 2ir^2e^{2i(\varphi-\psi)}$$
$$+ 2i\varepsilon\cos 3\psi + 2i\varepsilon r\cos 3\psi\cdot e^{i(\varphi-\psi)} + 3\varepsilon r^2\sin 3\psi - 3\varepsilon\sin 3\psi$$
$$- 6\varepsilon r\sin 3\psi\cdot e^{i(\varphi-\psi)} - 6\varepsilon r^2\sin 3\psi\cdot e^{2i(\varphi-\psi)} - 3i\varepsilon e^{3i\psi} - 4i\varepsilon r e^{i(\varphi+2\psi)}$$
$$+ 3i\varepsilon r^2 e^{3i\psi} - 2i\varepsilon r^2 e^{i(2\varphi+\psi)}\Big\}$$

$$= \frac{1}{2\pi}\int_0^{2\pi} f_2(\psi)\,d\psi\,\operatorname{Re}\Big\{r^2 - 1 - 2re^{i(\varphi-\psi)} - 2r^2e^{2i(\varphi-\psi)} + 2\varepsilon\cos 3\psi$$
$$+ 2\varepsilon r\cos 3\psi\cdot e^{i(\varphi-\psi)} - 3i\varepsilon r^2\sin 3\psi + 3i\varepsilon\sin 3\psi + 6i\varepsilon r\sin 3\psi\cdot e^{i(\varphi-\psi)}$$
$$+ 6i\varepsilon r^2\sin 3\psi\cdot e^{2i(\varphi-\psi)} - 3\varepsilon e^{3i\psi} - 4\varepsilon r e^{i(\varphi+2\psi)} + 3\varepsilon r^2 e^{3i\psi} - 2\varepsilon r^2 e^{i(2\varphi+\psi)}\Big\}.$$

Separating the real part, we find

$$V(r,\varphi) \approx \frac{1}{2\pi}\int_0^{2\pi} f_2(\psi)\,d\psi\,\Big\{r^2 - 1 - 2r\cos(\varphi-\psi) - 2r^2\cos 2(\varphi-\psi)$$
$$+ 2\varepsilon\cos 3\psi + 2\varepsilon r\cos 3\psi\,\cos(\varphi-\psi) - 6\varepsilon r\sin 3\psi\,\sin(\varphi-\psi)$$
$$- 6\varepsilon r^2\sin 3\psi\,\sin 2(\varphi-\psi) - 3\varepsilon\cos 3\psi - 4\varepsilon r\cos(\varphi+2\psi) + 3\varepsilon r^2\cos 3\psi$$
$$- 2\varepsilon r^2\cos(2\varphi+\psi)\Big\}$$

$$= \frac{1}{2\pi}\int_0^{2\pi} f_2(\psi)\,d\psi\,\Big\{r^2 - 1 - 2r\cos\varphi\,\cos\psi - 2r\sin\varphi\,\sin\psi$$
$$- 2r^2\cos 2\varphi\,\cos 2\psi - 2r^2\sin 2\varphi\,\sin 2\psi - \varepsilon\cos 3\psi + 2\varepsilon r\cos 3\psi\,\cos\varphi\,\cos\psi$$
$$+ 2\varepsilon r\cos 3\psi\,\sin\varphi\,\sin\psi - 6\varepsilon r\sin 3\psi\,\sin\varphi\,\cos\psi + 6\varepsilon r\sin 3\psi\,\cos\varphi\,\sin\psi$$
$$- 6\varepsilon r^2\sin 3\psi\,\sin 2\varphi\,\cos 2\psi + 6\varepsilon r^2\sin 3\psi\,\cos 2\varphi\,\sin 2\psi - 4\varepsilon r\cos\varphi\,\cos 2\psi$$
$$+ 4\varepsilon r\sin\varphi\,\sin 2\psi + 3\varepsilon r^2\cos 3\psi - 2\varepsilon r^2\cos 2\varphi\,\cos\psi + 2\varepsilon r^2\sin 2\varphi\,\sin\psi\Big\}$$

$$= \int_0^{2\pi}\frac{f_2(\psi)\,d\psi}{2\pi}\Big\{[1 + 3\varepsilon\cos 3\psi]r^2 + [-1 - \varepsilon\cos 3\psi]$$

$$+ r\cos\varphi\big[-2\cos\psi + 2\varepsilon\cos 3\psi\,\cos\psi + 6\varepsilon\sin 3\psi\,\sin\psi - 4\varepsilon\cos 2\psi\big]$$
$$+ r\sin\varphi\big[-2\sin\psi + 2\varepsilon\cos 3\psi\,\sin\psi - 6\varepsilon\sin 3\psi\,\cos\psi + 4\varepsilon\sin 2\psi\big]$$
$$+ r^2\cos 2\varphi\big[-2\cos 2\psi + 6\varepsilon\sin 3\psi\,\sin 2\psi - 2\varepsilon\cos\psi\big]$$
$$+ r^2\sin 2\varphi\big[-2\sin 2\psi - 6\varepsilon\sin 3\psi\,\cos 2\psi + 2\varepsilon\sin\psi\big]\Big\}.$$

In other words,

$$V(r,\varphi) = \bar{C}_0 r^2 - \bar{a}_0 - (\bar{a}_1\cos\varphi + \bar{b}_1\sin\varphi)r - (\bar{a}_2\cos 2\varphi + \bar{b}_2\sin 2\varphi)r^2 - \dots$$

where

$$\bar{a}_0 = \int_0^{2\pi} f_2(\psi)\psi_0(\psi)\,\mathrm{d}\psi, \qquad \bar{b}_1 = \int_0^{2\pi} f_2(\psi)\psi_2(\psi)\,\mathrm{d}\psi,$$
$$\bar{a}_1 = \int_0^{2\pi} f_2(\psi)\psi_1(\psi)\,\mathrm{d}\psi, \qquad \bar{b}_2 = \int_0^{2\pi} f_2(\psi)\psi_4(\psi)\,\mathrm{d}\psi,$$
$$\bar{a}_2 = \int_0^{2\pi} f_2(\psi)\psi_3(\psi)\,\mathrm{d}\psi, \qquad \bar{C}_0 = \int_0^{2\pi} f_2(\psi)\psi_5(\psi)\,\mathrm{d}\psi,$$

$$\psi_0(\psi) = \frac{1+\varepsilon\cos 3\psi}{2\pi},$$
$$\psi_1(\psi) = \frac{1}{2\pi}\big[2\cos\psi - 2\varepsilon\cos 3\psi\,\cos\psi - 6\varepsilon\sin 3\psi\,\sin\psi + 4\varepsilon\cos 2\psi\big],$$
$$\psi_2(\psi) = \frac{1}{2\pi}\big[2\sin\psi - 2\varepsilon\cos 3\psi\,\sin\psi + 6\varepsilon\sin 3\psi\,\cos\psi - 4\varepsilon\sin 2\psi\big],$$
$$\psi_3(\psi) = \frac{1}{2\pi}\big[2\cos 2\psi - 6\varepsilon\sin 3\psi\,\sin 2\psi + 2\varepsilon\cos\psi\big],$$
$$\psi_4(\psi) = \frac{1}{2\pi}\big[2\sin 2\psi + 6\varepsilon\sin 3\psi\,\cos 2\psi - 2\varepsilon\sin\psi\big],$$
$$\psi_5(\psi) = \frac{1+3\varepsilon\cos 3\psi}{2\pi}.$$

After simple transformations, we obtain

$$\psi_0(\psi) = \frac{1}{2\pi}[1+\varepsilon\cos 3\psi],$$
$$\psi_1(\psi) = \frac{1}{2\pi}[2\cos\psi + 2\varepsilon\cos 4\psi],$$

$$\psi_2(\psi) = \frac{1}{2\pi}[2\sin\psi + 2\varepsilon\sin 4\psi],$$

$$\psi_3(\psi) = \frac{1}{2\pi}[2\cos 2\psi - \varepsilon\cos\psi + 3\varepsilon\cos 5\psi],$$

$$\psi_4(\psi) = \frac{1}{2\pi}[2\sin 2\psi + \varepsilon\sin\psi + 3\varepsilon\sin 5\psi],$$

$$\psi_5(\psi) = \frac{1}{2\pi}[1 + 3\varepsilon\cos 3\psi]. \qquad (1.1.9)$$

Let us show that the Gram determinant of the system obtained is not equal to zero. To this end we find the integrals

$$a_{ij} = \int_0^{2\pi} \psi_i(\psi)\psi_j(\psi)\,\mathrm{d}\psi, \qquad i,j = \overline{0,5}. \qquad (1.1.10)$$

Taking into account that

$$\int_0^{2\pi} \sin m\psi\,\cos n\psi\,\mathrm{d}\psi = 0, \qquad m,n \in \mathbb{N},$$

$$\int_0^{2\pi} \sin m\psi\,\sin n\psi\,\mathrm{d}\psi = \int_0^{2\pi} \cos m\psi\,\cos n\psi\,\mathrm{d}\psi = 0, \quad m,n \in \mathbb{N}, \quad m \neq n,$$

we obtain

$$a_{00} = \frac{2+\varepsilon^2}{4\pi}, \qquad a_{05} = \frac{2+3\varepsilon^2}{4\pi} = a_{50}, \qquad a_{11} = \frac{1+\varepsilon^2}{\pi},$$

$$a_{13} = -\frac{\varepsilon}{2\pi} = a_{31}, \qquad a_{22} = \frac{1+\varepsilon^2}{\pi}, \qquad a_{24} = \frac{\varepsilon}{2\pi} = a_{42},$$

$$a_{33} = \frac{2+5\varepsilon^2}{2\pi}, \qquad a_{44} = \frac{2+5\varepsilon^2}{2\pi}, \qquad a_{55} = \frac{2+9\varepsilon^2}{4\pi}.$$

All other a_{ij} are equal to zero. The Gram determinant of the system $\{\psi_j\}_0^5$ is as follows:

$$\Delta = \begin{vmatrix} \frac{2+\varepsilon^2}{4\pi} & 0 & 0 & 0 & 0 & \frac{2+3\varepsilon^2}{4\pi} \\ 0 & \frac{1+\varepsilon^2}{\pi} & 0 & -\frac{\varepsilon}{2\pi} & 0 & 0 \\ 0 & 0 & \frac{1+\varepsilon^2}{\pi} & 0 & \frac{\varepsilon}{2\pi} & 0 \\ 0 & -\frac{\varepsilon}{2\pi} & 0 & \frac{2+5\varepsilon^2}{2\pi} & 0 & 0 \\ 0 & 0 & \frac{\varepsilon}{2\pi} & 0 & \frac{2+5\varepsilon^2}{2\pi} & 0 \\ \frac{2+3\varepsilon^2}{4\pi} & 0 & 0 & 0 & 0 & \frac{2+9\varepsilon^2}{4\pi} \end{vmatrix}$$

$$= \frac{2+\varepsilon^2}{4\pi} \cdot \frac{2+9\varepsilon^2}{4\pi} \widetilde{\Delta} - \frac{2+3\varepsilon^2}{4\pi} \cdot \frac{2+3\varepsilon^2}{4\pi} \widetilde{\Delta},$$

where $\widetilde{\Delta}$ is the central 4×4-minor

$$\widetilde{\Delta} = \begin{vmatrix} \frac{1+\varepsilon^2}{\pi} & 0 & -\frac{\varepsilon}{2\pi} & 0 \\ 0 & \frac{1+\varepsilon^2}{\pi} & 0 & \frac{\varepsilon}{2\pi} \\ -\frac{\varepsilon}{2\pi} & 0 & \frac{2+5\varepsilon^2}{2\pi} & 0 \\ 0 & \frac{\varepsilon}{2\pi} & 0 & \frac{2+5\varepsilon^2}{2\pi} \end{vmatrix}$$

$$= \frac{1+\varepsilon^2}{\pi} \begin{vmatrix} \frac{1+\varepsilon^2}{\pi} & 0 & \frac{\varepsilon}{2\pi} \\ 0 & \frac{2+5\varepsilon^2}{2\pi} & 0 \\ \frac{\varepsilon}{2\pi} & 0 & \frac{2+5\varepsilon^2}{2\pi} \end{vmatrix} - \frac{\varepsilon}{2\pi} \begin{vmatrix} 0 & \frac{1+\varepsilon^2}{\pi} & \frac{\varepsilon}{2\pi} \\ -\frac{\varepsilon}{2\pi} & 0 & 0 \\ 0 & \frac{\varepsilon}{2\pi} & \frac{2+5\varepsilon^2}{2\pi} \end{vmatrix}$$

$$= \frac{1+\varepsilon^2}{\pi} \Big[\frac{1+\varepsilon^2}{\pi} \Big(\frac{2+5\varepsilon^2}{2\pi} \Big)^2 - \frac{\varepsilon^2}{4\pi^2} \cdot \frac{2+5\varepsilon^2}{2\pi} \Big]$$

$$- \frac{\varepsilon}{2\pi} \Big[-\frac{\varepsilon^3}{8\pi^3} + \frac{\varepsilon}{2\pi} \cdot \frac{1+\varepsilon^2}{\pi} \cdot \frac{2+5\varepsilon^2}{2\pi} \Big]$$

$$= \frac{1}{16\pi^4}\Big[(2+2\varepsilon^2)^2(2+5\varepsilon^2)^2 - (2+2\varepsilon^2)\varepsilon^2(2+5\varepsilon^2) + \varepsilon^4 - \varepsilon^2(2+2\varepsilon^2)(2+5\varepsilon^2)\Big]$$

$$= \frac{1}{16\pi^4}\left[(2+2\varepsilon^2)(2+5\varepsilon^2) - \varepsilon^2\right]^2 = \frac{1}{16\pi^4}[4+4\varepsilon^2+10\varepsilon^2+10\varepsilon^4-\varepsilon^2]^2$$

$$= \frac{1}{16\pi^4}(4+13\varepsilon^2+10\varepsilon^4)^2.$$

Then

$$\Delta = \frac{\widetilde{\Delta}}{16\pi^2}\left(4+2\varepsilon^2+18\varepsilon^2+9\varepsilon^4-4-12\varepsilon^2-9\varepsilon^4\right) = \frac{\varepsilon^2}{2\pi^2}\widetilde{\Delta},$$

i. e., $\Delta \neq 0$ for $\varepsilon \neq 0$. Therefore, system (1.1.9) is linearly independent. Let us define the function $f_2(\psi)$ by the formula

$$f_2(\psi) = \sum_{i=0}^{5} D_i\psi_i(\psi). \tag{1.1.11}$$

Multiplying both sides of this equality by each of the functions $\{\psi_j\}_0^5$, integrating from 0 to 2π, and setting

$$\bar{C}_0 = \frac{\varepsilon^2}{r_1^2}, \qquad 0 < 2r_1 < 1-\varepsilon, \qquad \bar{a}_0 = \varepsilon^2,$$
$$\bar{a}_1 = \bar{b}_1 = \bar{a}_2 = \bar{b}_2 = 0,$$

we obtain

$$\int_0^{2\pi} f_2(\psi)\psi_0(\psi)\,d\psi = \sum_{i=0}^{5} a_{0i}D_i = \bar{a}_0 = \varepsilon^2,$$

$$\int_0^{2\pi} f_2(\psi)\psi_1(\psi)\,d\psi = \sum_{i=0}^{5} a_{1i}D_i = \bar{a}_1 = 0,$$

$$\int_0^{2\pi} f_2(\psi)\psi_2(\psi)\,d\psi = \sum_{i=0}^{5} a_{2i}D_i = \bar{b}_1 = 0,$$

$$\int_0^{2\pi} f_2(\psi)\psi_3(\psi)\,d\psi = \sum_{i=0}^{5} a_{3i}D_i = \bar{a}_2 = 0,$$

$$\int_0^{2\pi} f_2(\psi)\psi_4(\psi)\,d\psi = \sum_{i=0}^{5} a_{4i}D_i = \bar{b}_2 = 0,$$

$$\int_0^{2\pi} f_2(\psi)\psi_5(\psi)\,\mathrm{d}\psi = \sum_{i=0}^{5} a_{5i} D_i = \bar{C}_0 = \frac{\varepsilon^2}{r_1^2}. \tag{1.1.12}$$

Taking into account the linear independence of the system $\{\psi_j\}_0^5$, we find

$$D_i = \frac{\Delta_i}{\Delta}, \qquad i = \overline{0,5}, \tag{1.1.13}$$

where (the asterisk below stands for non-zero elements)

$$\Delta_0 = \begin{vmatrix} \varepsilon^2 & 0 & 0 & 0 & 0 & \frac{2+3\varepsilon^2}{4\pi} \\ 0 & \frac{1+\varepsilon^2}{\pi} & 0 & -\frac{\varepsilon}{2\pi} & 0 & 0 \\ 0 & 0 & \frac{1+\varepsilon^2}{\pi} & 0 & \frac{\varepsilon}{2\pi} & 0 \\ 0 & -\frac{\varepsilon}{2\pi} & 0 & \frac{2+5\varepsilon^2}{2\pi} & 0 & 0 \\ 0 & 0 & \frac{\varepsilon}{2\pi} & 0 & \frac{2+5\varepsilon^2}{2\pi} & 0 \\ \frac{\varepsilon^2}{r_1^2} & 0 & 0 & 0 & 0 & \frac{2+9\varepsilon^2}{4\pi} \end{vmatrix}$$

$$= \varepsilon^2 \cdot \frac{2+9\varepsilon^2}{4\pi}\widetilde{\Delta} - \frac{2+3\varepsilon^2}{4\pi}\cdot\frac{\varepsilon^2}{r_1^2}\widetilde{\Delta} = \frac{\widetilde{\Delta}\varepsilon^2}{4\pi}\left(2+9\varepsilon^2 - \frac{2+3\varepsilon^2}{r_1^2}\right),$$

$$\Delta_1 = \begin{vmatrix} * & * & 0 & 0 & 0 & * \\ 0 & 0 & 0 & * & 0 & 0 \\ 0 & 0 & * & 0 & * & 0 \\ 0 & 0 & 0 & * & 0 & 0 \\ 0 & 0 & * & 0 & * & 0 \\ * & * & 0 & 0 & 0 & * \end{vmatrix} = 0, \qquad \Delta_2 = \begin{vmatrix} * & 0 & * & 0 & 0 & * \\ 0 & * & 0 & * & 0 & 0 \\ 0 & 0 & 0 & 0 & * & 0 \\ 0 & * & 0 & * & 0 & 0 \\ 0 & 0 & 0 & 0 & * & 0 \\ * & 0 & * & 0 & 0 & * \end{vmatrix} = 0,$$

$$\Delta_3 = \begin{vmatrix} * & 0 & 0 & * & 0 & * \\ 0 & * & 0 & 0 & 0 & 0 \\ 0 & 0 & * & 0 & * & 0 \\ 0 & * & 0 & 0 & 0 & 0 \\ 0 & 0 & * & 0 & * & 0 \\ * & 0 & 0 & * & 0 & * \end{vmatrix} = 0, \qquad \Delta_4 = \begin{vmatrix} * & 0 & 0 & 0 & * & * \\ 0 & * & 0 & * & 0 & 0 \\ 0 & 0 & * & 0 & 0 & 0 \\ 0 & * & 0 & * & 0 & 0 \\ 0 & 0 & * & 0 & 0 & 0 \\ * & 0 & 0 & 0 & * & * \end{vmatrix} = 0,$$

$$\Delta_5 = \begin{vmatrix} \dfrac{2+\varepsilon^2}{4\pi} & 0 & 0 & 0 & 0 & \varepsilon^2 \\ 0 & \dfrac{1+\varepsilon^2}{\pi} & 0 & -\dfrac{\varepsilon}{2\pi} & 0 & 0 \\ 0 & 0 & \dfrac{1+\varepsilon^2}{\pi} & 0 & \dfrac{\varepsilon}{2\pi} & 0 \\ 0 & -\dfrac{\varepsilon}{2\pi} & 0 & \dfrac{2+5\varepsilon^2}{2\pi} & 0 & 0 \\ 0 & 0 & \dfrac{\varepsilon}{2\pi} & 0 & \dfrac{2+5\varepsilon^2}{2\pi} & 0 \\ \dfrac{2+3\varepsilon^2}{4\pi} & 0 & 0 & 0 & 0 & \dfrac{\varepsilon^2}{r_1^2} \end{vmatrix}$$

$$= \frac{2+\varepsilon^2}{4\pi}\cdot\frac{\varepsilon^2}{r_1^2}\widetilde{\Delta} - \varepsilon^2\cdot\frac{2+3\varepsilon^2}{4\pi}\widetilde{\Delta} = \frac{\widetilde{\Delta}\varepsilon^2}{4\pi}\Big(\frac{2+\varepsilon^2}{r_1^2} - 2 - 3\varepsilon^2\Big).$$

Hence it follows that

$$D_0 = \frac{\Delta_0}{\Delta} = \frac{\widetilde{\Delta}\varepsilon^2}{4\pi}\Big(2+9\varepsilon^2 - \frac{2+3\varepsilon^2}{r_1^2}\Big)\Big(\frac{\varepsilon^2\widetilde{\Delta}}{2\pi^2}\Big)^{-1} = \frac{\pi}{2}\Big(2+9\varepsilon^2 - \frac{2+3\varepsilon^2}{r_1^2}\Big),$$

$$D_1 = D_2 = D_3 = D_4 = 0,$$

$$D_5 = \frac{\Delta_5}{\Delta} = \frac{\widetilde{\Delta}\varepsilon^2}{4\pi}\Big(\frac{2+\varepsilon^2}{r_1^2} - 2 - 3\varepsilon^2\Big)\Big(\frac{\varepsilon^2\widetilde{\Delta}}{2\pi^2}\Big)^{-1} = \frac{\pi}{2}\Big(\frac{2+\varepsilon^2}{r_1^2} - 2 - 3\varepsilon^2\Big),$$

and

$$f_2(\psi) = \sum_{i=0}^{5} D_i\psi_i(\psi) = D_0\psi_0(\psi) + D_5\psi_5(\psi)$$

$$= \frac{\pi}{2}\Big(2+9\varepsilon^2 - \frac{2+3\varepsilon^2}{r_1^2}\Big)\frac{1+\varepsilon\cos 3\psi}{2\pi} + \frac{\pi}{2}\Big(\frac{2+\varepsilon^2}{r_1^2} - 2 - 3\varepsilon^2\Big)\frac{1+3\varepsilon\cos 3\psi}{2\pi}$$

$$= \frac{1}{4}\Big\{2+9\varepsilon^2 + \varepsilon(2+9\varepsilon^2)\cos 3\psi - 2 - 3\varepsilon^2 - 3\varepsilon(2+3\varepsilon^2)\cos 3\psi$$

$$+ \frac{1}{r_1^2}\Big(-2-3\varepsilon^2 - 2\varepsilon\cos 3\psi - 3\varepsilon^3\cos 3\psi + 2 + \varepsilon^2 + 6\varepsilon\cos 3\psi + 3\varepsilon^3\cos 3\psi\Big)\Big\}$$

$$= \frac{1}{4}\Big\{6\varepsilon^2 - 4\varepsilon\cos 3\psi + \frac{1}{r_1^2}(-2\varepsilon^2 + 4\varepsilon\cos 3\psi)\Big\}$$

$$= \frac{\varepsilon^2}{2}\Big(3 - \frac{1}{r_1^2}\Big) - \varepsilon\cos 3\psi\Big(1 - \frac{1}{r_1^2}\Big) = -\varepsilon\cos 3\psi\Big(1 - \frac{1}{r_1^2}\Big) + O(\varepsilon^2). \quad (1.1.14)$$

In this case, (1.1.4) will be as follows

$$V(r,\varphi) = \varepsilon^2\left(\frac{r}{r_1}\right)^2 - \varepsilon^2 + F(\varepsilon, r, \varphi), \tag{1.1.15}$$

where F is some function biharmonic in D_1. We have

$$V(0,\varphi) = -\varepsilon^2, \qquad V(2r_1,\varphi) = 3\varepsilon^2 + F(\varepsilon, 2r_1, \varphi), \qquad F(\varepsilon, 2r_1, \varphi) = O(r_1^3).$$

Further, we obtain

$$V_r'(r,\varphi) = 2r\left(\frac{\varepsilon}{r_1}\right)^2 + r^2 F_1(\varepsilon, r, \varphi), \qquad F_1(\varepsilon, r, \varphi) = O(1).$$

Hence it follows that

$$V(0,\varphi) < 0, \qquad V(2r_1,\varphi) > 0$$

and for a sufficiently small r_1 we have

$$V_r'(r,\varphi) > 0, \qquad r < 2r_1.$$

Therefore, there exists a closed contour $r = R_2(\varphi)$ lying inside the circle $(0, 2r_1)$ such that

$$V(R_2(\varphi),\varphi) = 0.$$

So, (1.1.15) is a nontrivial solution of equation (1) for homogeneous conditions corresponding to (2). Note that the system $\{\psi_j\}_0^5$ is linearly dependent for $\varepsilon = 0$; i. e., the Gram determinant of this system is equal to zero. Therefore, in such a way we can construct a counterexample only for such contours S_1 whose systems of functions (1.1.9) are linearly independent.

1.2. THE FIRST INTERNAL BOUNDARY VALUE PROBLEM IN A DISK

Definition. We shall call a problem *completely solvable* if it is solvable on a complete set.

Theorem 1.2.1. *If one of the contours S_j, $j = 1, 2$, is a circle, then the problem* (1), (2) *is uniquely completely solvable.*

Proof. To prove the theorem we use the representation

$$u(r,\varphi) = u_1(r,\varphi) + r^2 u_2(r,\varphi), \tag{1.2.1}$$

where

$$u_j(p) = \int_{S_1} \mu_j(q) \frac{\partial}{\partial n_q} G(p,q) \, \mathrm{d}_q S_1, \quad j = 1,2.$$

Then, instead of system (1.0.2), we have

$$\begin{aligned} &\mu_1(p) + R_1^2(p)\mu_2(p) = \Phi_1(p), \quad p \in S_1, \\ &\int_{S_1} \mu_1(q) \frac{\partial}{\partial n_q} G(p,q) \, \mathrm{d}_q S_1 + R_1^2(p) \int_{S_1} \mu_2(q) \frac{\partial}{\partial n_q} G(p,q) \, \mathrm{d}_q S_1 \\ &\qquad = \Phi_2(p), \quad p \in S_2, \end{aligned} \tag{1.2.2}$$

where $r = R_j(p)$ are the equations for S_j, $j = 1,2$.

Now, to prove the theorem it suffices to establish the unique complete solvability of system (1.2.2). Uniqueness under the conditions of the theorem follows from (1.2.2).

As is known, system (1.2.2) is completely solvable if and only if the solutions of the associated homogeneous system of Fredholm integral equations of third kind

$$\begin{aligned} &\nu_1(q) + \int_{S_2} \nu_2(p) \frac{\partial}{\partial n_q} G(p,q) \, \mathrm{d}_p S_2 = 0, \quad q \in S_1, \\ &R_1^2(q)\nu_1(q) + \int_{S_2} R_2^2(p)\nu_2(p) \frac{\partial}{\partial n_q} G(p,q) \, \mathrm{d}_p S_2 = 0, \quad q \in S_1, \end{aligned} \tag{1.2.3}$$

are trivial. Hence, when S_1 is a circle, we have

$$\int_{S_2} \nu_2(p)[R_1^2 - R_2^2(p)] \frac{\partial}{\partial n_q} G(p,q) \, \mathrm{d}_p S_2 = 0, \quad q \in S_1. \tag{1.2.4}$$

Now consider the function

$$V_2(q) = \int_{S_2} \nu_2(p)[R_1^2 - R_2^2(p)] G(p,q) \, \mathrm{d}_p S_2$$

for $q \in D_1$. It is evident that

$$\Delta V_2(q) = 0, \quad q \in D_1,$$

$$V_2(q) = 0, \quad \frac{\partial}{\partial n_q} V_2(q) = 0, \quad q \in S_1.$$

Then, by uniqueness of the solution of the Cauchy problem for the Laplace equation in $D_1 \setminus \overline{D_2}$, we find

$$V_2(q) = 0, \quad q \in D_1 \setminus \overline{D_2}.$$

By the property of limit passage of $V_2(q)$ across S_2 (Gyunter, 1953, p. 134), we get

$$V_2(q) = 0, \quad q \in D_2.$$

Therefore, the equality

$$\nu_2(q) = 0, \quad q \in S_2, \quad \text{in } L(S_2)$$

holds at all the Lebesgue points q (Gyunter, 1953, p. 132) of the integrable function $\nu_2(q)$, since

$$\Big(\frac{\partial}{\partial n_q} V_2(q)\Big)_e - \Big(\frac{\partial}{\partial n_q} V_2(q)\Big)_i = \pi \nu_2(q)[R_1^2 - R_2^2(q)], \quad q \in S_2.$$

The equality

$$\nu_1(q) = 0, \quad q \in S_1,$$

follows from (1.2.3).

We suppose now that S_2 is a circle. Then (1.2.3) yields

$$[R_1^2(q) - R_2^2] \int_{S_2} \nu_2(p) \frac{\partial}{\partial n_q} G(p,q)\, \mathrm{d}_p S_2 = 0, \quad q \in S_1. \tag{1.2.5}$$

Since $R_1^2(q) > R_2^2$, (1.2.5) yields

$$\int_{S_2} \nu_2(p) \frac{\partial}{\partial n_q} G(p,q)\, \mathrm{d}_p S_2 = 0, \quad q \in S_1.$$

Hence, analogously to (1.2.4), we obtain

$$\nu_j(p) = 0, \quad p \in S_j, \quad j = 1, 2.$$

□

If S_1 and S_2 are circles, then a theorem characterizing the conditional well-posedness of the problem (1), (2) holds.

Theorem 1.2.2. *If the considered function $u(r, \varphi)$ satisfies the inequalities*

$$\frac{1}{\pi} \int_{-\pi}^{\pi} u^2(1,\varphi)\, \mathrm{d}\varphi \le \varepsilon_1, \qquad \frac{1}{\pi} \int_{-\pi}^{\pi} u^2(\rho,\varphi)\, \mathrm{d}\varphi \le \varepsilon_2, \quad \rho < 1,$$

$$\frac{1}{\pi}\int_{-\pi}^{\pi} u_2^2(1,\varphi)\,\mathrm{d}\varphi \le 1, \tag{1.2.6}$$

then the following inequality also holds:

$$\frac{1}{\pi}\int_{-\pi}^{\pi} u_2^2(r,\varphi)\,\mathrm{d}\varphi \le 2\Big\{\varepsilon_2 + \exp\Big[\frac{\ln r}{\ln\rho}\ln\frac{2(\varepsilon_1^2+\varepsilon_2^2)}{1-\rho^2}\Big]\Big\}, \tag{1.2.7}$$

where $u_2(r,\varphi) = \dfrac{1}{4r}\displaystyle\int_0^r \Delta u\,\mathrm{d}r.$

Proof. From representation (1.0.3), taking into account the expansion

$$u_j(r,\varphi) = \frac{a_0^{(j)}}{2} + \sum_{n=1}^{\infty} r^n(a_n^{(j)}\cos n\varphi + b_n^{(j)}\sin n\varphi), \quad j = 1,2,$$

and (1.2.6), we get

$$\frac{1}{\pi}\int_{-\pi}^{\pi} u_1^2(r,\varphi)\,\mathrm{d}\varphi \le \varepsilon_1,$$

$$\frac{1}{\pi}\int_{-\pi}^{\pi} u_2^2(\rho,\varphi)\,\mathrm{d}\varphi \le \frac{2(\varepsilon_1+\varepsilon_2)}{(1-\rho^2)^2}.$$

In other words, we have

$$\sum_{n=0}^{\infty}\rho^{2n}D_n^2 \le \frac{2(\varepsilon_1+\varepsilon_2)}{(1-\rho^2)^2}, \qquad \sum_{n=0}^{\infty} D_n^2 \le 1, \tag{1.2.8}$$

where

$$D_n = \sqrt{(a_n^{(2)})^2 + (b_n^{(2)})^2}, \qquad D_0 = \frac{a_0^{(2)}}{\sqrt{2}}.$$

Taking into account (1.2.8), we estimate the function

$$w(r) = \sum_{n=0}^{\infty} r^{2n}D_n^2 = \frac{1}{\pi}\int_{-\pi}^{\pi} u_2^2(r,\varphi)\,\mathrm{d}\varphi. \tag{1.2.9}$$

By (1.2.8), this sum is bounded. Suppose that the sum (1.2.9) achieves its conditional maximum with respect to the variables D_n for $D_n = \bar{D}_n$ $(n = \overline{0,\infty}\,)$.

It is easy to see that

$$\bar{D}_n = 0, \quad n \ne p, q,$$

where p, q $(p < q)$ are some numbers.

By similar considerations, $\bar{D}_p$ and $\bar{D}_q$ satisfy one of the following relations

$$\begin{aligned} &\text{(I)} && \bar{D}_p^2 + \bar{D}_q^2 = 1, \quad \bar{D}_p^2 \rho^{2p} + \bar{D}_q^2 \rho^{2q} = 2(1-\rho^2)^{-2}(\varepsilon_1 + \varepsilon_2); \\ &\text{(II)} && \bar{D}_p = 0; \\ &\text{(III)} && \bar{D}_q = 0. \end{aligned} \tag{1.2.10}$$

Suppose that the first relation from (1.2.10) holds. Then

$$\begin{aligned} \bar{D}_p^2 &= \frac{2(1-\rho^2)^{-2}(\varepsilon_1+\varepsilon_2)}{\rho^{2p}-\rho^{2q}} - \frac{\rho^{2q}}{\rho^{2p}-\rho^{2q}}, \\ \bar{D}_q^2 &= \frac{\rho^{2p}}{\rho^{2p}-\rho^{2q}} - \frac{2(1-\rho^2)^{-2}(\varepsilon_1+\varepsilon_2)}{\rho^{2p}-\rho^{2q}}. \end{aligned} \tag{1.2.11}$$

Note that (1.2.11) implies

$$\rho^{2q} \le 2(1-\rho^2)^{-2}(\varepsilon_1+\varepsilon_2) \le \rho^{2p}. \tag{1.2.12}$$

Taking into account (1.2.10) and (1.2.11), we obtain

$$\sum_{n=0}^{\infty} \bar{D}_n^2 r^{2n} = \frac{2(1-\rho^2)^{-2}(\varepsilon_1+\varepsilon_2)(r^{2p}-r^{2q})}{\rho^{2p}-\rho^{2q}} + \frac{\rho^{2p}r^{2q}-\rho^{2q}r^{2p}}{\rho^{2p}-\rho^{2q}}. \tag{1.2.13}$$

It is easy to show that the functions

$$F_1(p,q) = \frac{(1-\rho^2)^{-2}(r^{2p}-r^{2q})}{\rho^{2p}-\rho^{2q}}, \qquad F_2(p,q) = \frac{\rho^{2p}r^{2q}-\rho^{2q}r^{2p}}{\rho^{2p}-\rho^{2q}} \quad (p<q)$$

increase when p increases and q decreases. Therefore, taking into account (1.2.12) and (1.2.13), we have

$$\begin{aligned} \sum_{n=0}^{\infty} \bar{D}_n^2 r^{2n} &\le \lim_{p,q\to\alpha} \left[2(\varepsilon_1+\varepsilon_2)F_1(p,q) + F_2(p,q)\right] \\ &= \exp\left[\frac{\ln r}{\ln \rho} \ln \frac{2(\varepsilon_1+\varepsilon_2)}{(1-\rho^2)^2}\right], \\ \alpha &= \frac{1}{2\ln\rho} \ln \frac{2(\varepsilon_1+\varepsilon_2)}{(1-\rho^2)^2}. \end{aligned} \tag{1.2.14}$$

Suppose now that the second relation from (1.2.10) holds. Then

$$\sum_{n=0}^{\infty} \bar{D}_n^2 r^{2n} = \bar{D}_q^2 r^{2q}$$

and (1.2.6) yields

$$\bar{D}_q^2 \le 1, \qquad \bar{D}_q^2 r^{2q} \le \frac{2(\varepsilon_1 + \varepsilon_2)}{(1-\rho^2)^2}.$$

Hence we obtain

$$\sum_{n=0}^{\infty} \bar{D}_n^2 r^{2n} \le \exp\Big\{\frac{\ln r}{\ln \rho} \ln \frac{2(\varepsilon_1 + \varepsilon_2)}{(1-\rho^2)^2}\Big\}. \tag{1.2.15}$$

If the third relation from (1.2.10) holds, inequality (1.2.15) is proved analogously. Then the theorem follows from (1.0.3), (1.2.14), (1.2.15), and the definition of the numbers $\bar{D}_n$. □

Consideration of an essentially ill-posed problem as a conditionally well-posed one gives us the possibility to construct an approximate solution. Given approximate data, this solution can be constructed with guaranteed accuracy, for example, by the regularization method (Lavrent'ev, 1962, p. 25; Tikhonov and Arsenin, 1977).

Really, consider the family of linear operators $B_n^{(j)}$ depending on the integer parameter n which are defined as follows

$$B_n^{(1)}\Phi_1(\varphi) = \sum_{k=1}^{n} r^k (a_k^{(1)} \cos k\varphi + b_k^{(1)} \sin k\varphi) + \frac{a_0^{(1)}}{2},$$

$$B_n^{(2)}\Phi_2(\varphi) = (1-r^2)\left[\sum_{k=1}^{n} r^k (a_k^{(2)} \cos k\varphi + b_k^{(2)} \sin k\varphi) + \frac{a_0^{(2)}}{2}\right], \tag{1.2.16}$$

$$a_k^{(1)} = \frac{1}{\pi}\int_{-\pi}^{\pi} \Phi_1(\varphi)\cos k\varphi \, \mathrm{d}\varphi, \qquad (1-\rho^2)a_k^{(2)} = \frac{1}{\pi}\int_{-\pi}^{\pi} \Phi_2(\varphi)\cos k\varphi \, \mathrm{d}\varphi,$$

$$b_k^{(1)} = \frac{1}{\pi}\int_{-\pi}^{\pi} \Phi_1(\varphi)\sin k\varphi \, \mathrm{d}\varphi, \qquad (1-\rho^2)b_k^{(2)} = \frac{1}{\pi}\int_{-\pi}^{\pi} \Phi_2(\varphi)\sin k\varphi \, \mathrm{d}\varphi,$$

$$k = \overline{0, n}.$$

It is easy to see that the family of operators $B_n^{(j)}$ will be a regularizing family if we consider the data Φ_1, Φ_2 and the solution $u(r,\varphi)$ as elements of the Hilbert space $L_2(-\pi,\pi)$ for each $r \in [0,1]$. Indeed, the operators of determination of the coefficients $a_k^{(j)}$ and $b_k^{(j)}$, $j = 1,2$, and of finite summation are continuous. Hence it follows that the operators $B_n^{(j)}$ are continuous (for each $n < \infty$ the problem is classically well-posed). Convergence of the sequence $B_n^{(1)}\Phi_1(\varphi) + B_n^{(2)}\Phi_2(\varphi)$ to the solution $u(r, \varphi)$ for $n \to \infty$ follows from representation (1.0.3) of the solution.

Now let us estimate the efficiency of application of the above regularizing family to construction of an approximate solution by approximate data. We assume that our problem is set conditionally well-posed and the set of well-posedness is determined by the inequality

$$\frac{1}{\pi}\int_{-\pi}^{\pi}\left[\frac{1}{4}\int_0^1 \Delta u \, \mathrm{d}r\right]^2 \mathrm{d}\varphi \leq 1.$$

Evidently, $\|B_n^{(1)}\| = r^n$, $\|B_n^{(2)}\| = (r/\rho)^n$. Thus, we need to estimate

$$\begin{aligned} &\|B_n^{(1)}\Phi_1 + B_n^{(2)}\Phi_2 - u(r,\varphi)\|^2 \\ &\leq \sum_{k=n+1}^{\infty}\left\{\left[r^k a_k^{(1)} + (1-r^2)\left(\frac{r}{\rho}\right)^k a_k^{(2)}\right]^2 + \left[r^k b_k^{(1)} + (1-r^2)\left(\frac{r}{\rho}\right)^k b_k^{(2)}\right]^2\right\} \\ &\leq \sum_{k=n+1}^{\infty}\left[r^{2k}\left([a_k^{(1)}]^2 + [b_k^{(1)}]^2\right) + (1-r^2)^2\left(\frac{r}{\rho}\right)^{2k} D_k^2\right] \end{aligned} \tag{1.2.17}$$

under the condition

$$\sum_{k=1}^{\infty} D_k^2 \rho^{-2k} \leq 1. \tag{1.2.18}$$

It is easy to see that the sum on the right-hand side of (1.2.17) achieves its maximal value under condition (1.2.18) in the case when

$$D_k = 0, \quad k \neq n+1, \qquad D_{n+1} = \rho^{2(n+1)}.$$

Therefore,

$$\|B_n^{(1)}\Phi_1 + B_n^{(2)}\Phi_2 - u(r,\varphi)\|^2 \leq [\varepsilon_1^2 + (1-r^2)^2]r^{2(n+1)}.$$

Suppose that $\Phi_1(\varphi)$ and $\Phi_2(\varphi)$ are known with accuracy ε; i. e., we know $\Phi_{1\varepsilon}(\varphi)$ and $\Phi_{2\varepsilon}(\varphi)$ such that

$$\|\Phi_j - \Phi_{j\varepsilon}\| \leq \varepsilon, \quad j = 1, 2.$$

In this case, as an approximate solution we take

$$u_{n\varepsilon} = B_n^{(1)}\Phi_1 + B_n^{(2)}\Phi_2$$

and estimate the difference between $u_{n\varepsilon}$ and the exact solution u:

$$\|u_{n\varepsilon} - u\| \le \|B_n^{(1)}\|\varepsilon_1 + \|B_n^{(2)}\|\varepsilon_2 + \gamma(u, n),$$

where

$$\gamma(u, n) = r^{n+1}\varepsilon_1 + (1 - r^2)r^{n+1}.$$

Note that the efficiency of regularization now depends on the choice of the regularization parameter n. For example, from the exponential equation

$$r^n\varepsilon_1 + \left(\frac{r}{\rho}\right)^n \varepsilon_2 = r^{n+1}\varepsilon_1 + (1 - r^2)r^{n+1}$$

we can find n.

Theorem 1.2.3. *If the function in question satisfies the inequalities*

$$\frac{1}{\pi}\int_{-\pi}^{\pi} u^2(1, \varphi)\,\mathrm{d}\varphi \le \varepsilon_1, \qquad \frac{1}{\pi}\int_{-\pi}^{\pi} u^2(\rho, \varphi)\,\mathrm{d}\varphi \le \varepsilon_2, \quad \rho < 1,$$

$$\frac{1}{\pi}\int_{-\pi}^{\pi} \left[\frac{\partial^k u_2(r, \varphi)}{\partial \varphi^k}\right]_{r=1}^2 \mathrm{d}\varphi \le 1, \tag{1.2.19}$$

then the following inequality also holds:

$$\frac{1}{\pi}\int_{-\pi}^{\pi} u^2(r, \varphi)\,\mathrm{d}\varphi \le \varepsilon_1 + \frac{1}{\lambda^{2k}}, \tag{1.2.20}$$

where

$$u_2(r, \varphi) = \frac{1}{4r}\int_0^r \Delta u\,\mathrm{d}\tau,$$

and λ is the root of the equation

$$2\lambda \ln \rho - 2k \ln \lambda = \ln \frac{\varepsilon_1 + \varepsilon_2}{(1 - \rho^2)^2}. \tag{1.2.21}$$

Note that for $(\varepsilon_1 + \varepsilon_2)/(1 - \rho^2)^2$ sufficiently small the root of equation (1.2.21) satisfies the inequality

$$\lambda > \frac{1}{4 \ln \rho} \ln \frac{\varepsilon_1 + \varepsilon_2}{(1 - \rho^2)^2}.$$

Theorem 1.2.4. *Suppose that*

$$\frac{1}{\pi}\int_{-\pi}^{\pi} u^2(1,\varphi)\,d\varphi \le \varepsilon_1, \qquad \frac{1}{\pi}\int_{-\pi}^{\pi} u^2(\rho,\varphi)\,d\varphi \le \varepsilon_2, \quad \rho < 1,$$

$$\frac{1}{\pi}\int_{-\pi}^{\pi} \left| r\frac{\partial u_2}{\partial r}\right|^2_{r=1} d\varphi \le 1. \tag{1.2.22}$$

Then

$$\frac{1}{\pi}\int_{-\pi}^{\pi} u^2(r,\varphi)\,d\varphi \le \varepsilon_1 + \frac{r^q}{\lambda^2},$$

where q is the root of the equation

$$2q\ln\rho - 2\ln\lambda = \ln\frac{\varepsilon_1+\varepsilon_2}{(1-\rho^2)^2}, \qquad \lambda = \mathrm{const} > 0.$$

The proofs of the last two theorems are similar to that of Theorem 1.2.2.

Remark 1.2.1. Theorems 1.2.2–1.2.4 also hold for an n-dimensional ball.

Remark 1.2.2. If S_2 is an arbitrary Lyapunov contour, then solving at first the Dirichlet problem for the Laplace equation and taking the value of this solution on a circle lying inside D_2 we obtain the case considered above.

Remark 1.2.3. If S_1 is an arbitrary Lyapunov star-like contour and S_2 is a circle, then, by means of the representation

$$u(r,\varphi) = u_1(r,\varphi) + (R_2^2 - r^2)u_2(r,\varphi),$$

we can reduce the problem to the problem which was already studied.

1.3. THE FIRST INTERNAL BOUNDARY VALUE PROBLEM IN A HALF-STRIP

In the half-strip $D = \{0 < x < \pi,\ y > 0\}$, we consider IBVP 1 in the following special setting. We need to find the solution of the biharmonic equation in D satisfying the conditions

$$u(x,0) = \varphi_1(x), \qquad u(x,b) = \varphi_2(x), \quad b > 0,$$

$$u(0,y) = u(\pi,y) = \Delta u(0,y) = \Delta u(\pi,y) = 0, \qquad u \to 0 \quad \text{for} \quad y \to \infty.$$

In D, a biharmonic function is as follows (Tikhonov and Samarskii, 1990):

$$u = u_1 + yu_2, \tag{1.3.1}$$

where u_j, $j = 1, 2$, are harmonic functions. By (1.3.1), taking into account the problem conditions, we obtain the following problems

$$\Delta u_1(x, y) = 0, \qquad (x, y) \in D, \tag{1.3.2}$$

$$u_1(x, 0) = \varphi_1(x), \qquad x \in (0, \pi), \tag{1.3.3}$$

$$u_1(0, y) = u_1(\pi, y) = 0, \qquad u_1 \to 0 \quad \text{for} \quad y \to \infty, \tag{1.3.4}$$

$$\Delta u_2(x, y) = 0, \qquad (x, y) \in D, \tag{1.3.5}$$

$$u_2(x, b) = \varphi_3(x), \qquad x \in (0, \pi), \tag{1.3.6}$$

$$u_2(0, y) = u_2(\pi, y) = 0, \qquad u_2 \to 0 \quad \text{for} \quad y \to \infty, \tag{1.3.7}$$

where

$$u_2 = -\frac{1}{2}\int_0^{b-y} \Delta u \, \mathrm{d}\tau, \qquad \varphi_3(x) = -\frac{1}{b}\, u_1(x, b) + \frac{1}{b}\, \varphi_2(x).$$

Evidently, the problem (1.3.2)–(1.3.4) for $\varphi_1(x) \in L_2(0, \pi)$ is well-posed in classical sense (the Dirichlet problem for the Laplace equation in D) and has the solution

$$u_1(x, y) = \sum_{k=1}^{\infty} a_k \mathrm{e}^{-ky} \sin kx, \tag{1.3.8}$$

where

$$a_k = \frac{2}{\pi}\int_0^{\pi} \varphi_1(x) \sin kx \, \mathrm{d}x.$$

However, the problem (1.3.5)–(1.3.7) (the problem of harmonic continuation) (see Lavrent'ev, 1973, p. 9) for $\varphi_3(x) \in L_2(0, \pi)$ is ill-posed. Indeed, the function

$$u_2(x, y) = \alpha \mathrm{e}^{-k(y-b)} \sin kx$$

solves the problem (1.3.5)–(1.3.7) for $\varphi_3(x) = \alpha \sin kx$. It is easy to see that for each pair of spaces of functions of the variable x and for any constants $\varepsilon > 0$, $C > 0$, we can choose α and k such that the inequalities

$$\|\alpha \sin kx\| \leq \varepsilon, \qquad \|\alpha \mathrm{e}^{-k(y-b)} \sin kx\| > C$$

hold. Hence it follows that the considered problem of harmonic continuation is set ill-posedly in the classical sense.

Let us find out whether the problem is conditionally well-posed. Firstly, we establish a theorem characterizing the stability estimates of the setting in $L_2(0,\pi)$.

Theorem 1.3.1. *If $u(0,y) = u(\pi,y) = \Delta u(0,y) = \Delta u(\pi,y) = 0$, $u \to 0$ for $y \to \infty$, and*

$$\frac{2}{\pi}\int_0^\pi u^2(x,0)\,\mathrm{d}x = \frac{2}{\pi}\int_0^\pi \varphi_1^2(x)\,\mathrm{d}x \le \varepsilon_1, \tag{1.3.9}$$

$$\frac{2}{\pi}\int_0^\pi u^2(x,b)\,\mathrm{d}x = \frac{2}{\pi}\int_0^\pi \varphi_2^2(x)\,\mathrm{d}x \le \varepsilon_2, \tag{1.3.10}$$

$$\frac{2}{\pi}\int_0^\pi u_2^2(x,0)\,\mathrm{d}x \le M, \qquad u_2(x,y) = -\frac{1}{2}\int_0^{b-y} \Delta u\,\mathrm{d}\tau, \tag{1.3.11}$$

then

$$\frac{2}{\pi}\int_0^\pi u^2(x,y)\,\mathrm{d}x \le 2\Big[\varepsilon_1 + y^2\Big(\frac{4\varepsilon_2}{b^2}\Big)^{y/b} M^{(b-y)/b}\Big], \quad 0 < y < \infty. \tag{1.3.12}$$

Note that inequality (1.3.12) implies uniqueness of solution of the problem in question and conditional well-posedness in the class (1.3.11).

Proof. Taking into account (1.3.9), we obtain from (1.3.8)

$$\frac{2}{\pi}\int_0^\pi u_1^2(x,0)\,\mathrm{d}x = \sum_{k=1}^\infty a_k^2 \le \varepsilon_1, \qquad \frac{2}{\pi}\int_0^\pi u_1^2(x,y)\,\mathrm{d}x = \sum_{k=1}^\infty a_k^2 \mathrm{e}^{-2ky}.$$

Hence we have

$$\frac{2}{\pi}\int_0^\pi u_1^2(x,y)\,\mathrm{d}x \le \varepsilon_1. \tag{1.3.13}$$

Further, (1.3.1) for $y = b$ and (1.3.10) and (1.3.13) yield

$$\frac{2}{\pi}\int_0^\pi u_2^2(x,b)\,\mathrm{d}x \le \frac{4\varepsilon_2}{b^2}. \tag{1.3.14}$$

Let $f(y)$ denote the function (see Lavrent'ev, 1981)

$$f(y) = \int_0^\pi u_2^2(x,y)\,\mathrm{d}x.$$

Evidently, $f(y) \to 0$ for $y \to \infty$. Differentiating $f(y)$, we obtain

$$f'(y) = 2\int_0^\pi u_2 u'_{2y}\,\mathrm{d}x,$$

$$f''(y) = 2\int_0^\pi [u'_{2y}]^2\,\mathrm{d}x + 2\int_0^\pi u_2 u''_{2yy}\,\mathrm{d}x.$$

Now transform the second term in the expression for $f''(y)$

$$\int_0^\pi u_2 u''_{2yy}\,\mathrm{d}x = -\int_0^\pi u_2 u''_{2xx}\,\mathrm{d}x = \int_0^\pi [u'_{2x}]^2\,\mathrm{d}x,$$

$$\frac{\mathrm{d}}{\mathrm{d}y}\int_0^\pi [u'_{2x}]^2\,\mathrm{d}x = 2\int_0^\pi u'_2 u''_{2xy}\,\mathrm{d}x = -2\int_0^\pi u'_{2y} u''_{2x}\,\mathrm{d}x$$

$$= 2\int_0^\pi u'_{2y} u''_{2yy}\,\mathrm{d}x = \frac{\mathrm{d}}{\mathrm{d}y}\int_0^\pi [u'_{2y}]^2\,\mathrm{d}x.$$

Therefore

$$f''(y) = 4\int_0^\pi [u'_{2y}]^2\,\mathrm{d}x.$$

Now consider the function

$$\varphi(y) = \ln f(y).$$

Differentiating this function we obtain

$$\varphi''(y) = \frac{1}{f^2(y)}\left[f''(y)f(y) - f'^2(y)\right]$$

$$= \frac{1}{f^2(y)}\left[4\int_0^\pi u_2^2\,\mathrm{d}x\int_0^\pi [u'_{2y}]^2\,\mathrm{d}x - \left(2\int_0^\pi u_2 u'_{2y}\,\mathrm{d}x\right)^2\right] \geq 0 \tag{1.3.15}$$

(by the well-known Cauchy – Schwartz – Bunyakovskii inequality). Inequality (1.3.15) means that the function $\varphi(y)$ is concave. Hence the function $\varphi(y)$ satisfies the inequality

$$\varphi(y) \leq \frac{b-y}{b}\,\varphi(0) + \frac{y}{b}\,\varphi(b). \tag{1.3.16}$$

Exponentiating this inequality, we obtain

$$f(y) \leq [f(b)]^{y/b}[f(0)]^{(b-y)/b}$$

or

$$\frac{2}{\pi}\int_0^\pi u_2^2(x,y)\,\mathrm{d}x \leq \left[\frac{4\varepsilon_2}{b^2}\right]^{y/b} M^{(b-y)/b}, \qquad \rho \leq y \leq \infty. \tag{1.3.17}$$

Formula (1.3.1) and the inequality

$$2yu_1u_2 \le u_1^2 + y^2u_2^2$$

yield

$$u^2 \le 2(u_1^2 + y^2u_2^2).$$

This inequality, (1.3.14), and (1.3.17) yield inequality (1.3.12). □

Consideration of an ill-posed problem in classical sense as a conditionally well-posed problem allows us to construct an approximate solution with guaranteed accuracy by approximate data similar as in Section 1.2:

$$\|u - u_{n\varepsilon}\| \le \varepsilon_1 + \frac{1}{n}\,\mathrm{e}^{nb}\varepsilon_2 + \gamma(u,n), \tag{1.3.18}$$

where

$$\gamma(u,n) = \mathrm{e}^{-(n+1)y}\sqrt{2\varepsilon_1 + y^2M}.$$

From the transcendental equation

$$\varepsilon_1 + \frac{1}{n}\,\mathrm{e}^{nb}\varepsilon_2 = \mathrm{e}^{-(n+1)y}\sqrt{2\varepsilon_1 + y^2M}$$

we find the regularization parameter n.

Suppose now that $D_1 = \{-\pi < x < \pi,\ y > 0\}$ and consider the problem of regularization in the uniform metric in the class of functions periodic relative to x. The condition in this case will be weakened, namely, we require only existence of the integral

$$-\frac{1}{2}\int_0^{b-y} \Delta u\,\mathrm{d}\tau \qquad \text{for } y = 0.$$

Let us take u_1 in the form

$$u_1(x,y) = \sum_{n=-\infty}^{\infty} C_{1n}\mathrm{e}^{-ny+inx}, \qquad C_{1n} = \frac{1}{2\pi}\int_{-\pi}^{\pi} \varphi_1(x)\mathrm{e}^{-inx}\,\mathrm{d}x, \tag{1.3.19}$$

and consider the function

$$u_{2\alpha}(x,y) = \frac{1}{4\pi^2}\int_{-\pi}^{\pi} \varphi_3(\tau)R_\alpha(x-\tau,y)\,\mathrm{d}\tau, \tag{1.3.20}$$

where

$$R_\alpha(x-\tau,y) = \sum_{n=-\infty}^{\infty} \mathrm{e}^{-\alpha^2n^2-n(y-b)}\int_{-\pi}^{\pi} \mathrm{e}^{\mathrm{i}n(x-\tau-\xi)}\,\mathrm{d}\xi.$$

Let

$$u_2(x,0) = \varphi(x), \qquad u_{2\alpha}(x,0) = \varphi_\alpha(x).$$

Under the conditions stated above, the function $\varphi(x)$ exists and is continuous.

Theorem 1.3.2. *Let $\omega(h)$ be the continuity modulus of the function $\varphi(x)$. Then*

$$|\varphi(x) - \varphi_\alpha(x)| \le \omega(\alpha)\Big[1 + \frac{2}{\sqrt{\pi}} + K(\alpha)\Big], \tag{1.3.21}$$

where $K(\alpha) \to 0$ for $\alpha \to 0$.

Proof. It is easy to see that

$$\frac{1}{\pi}\Big[\frac{1}{2} + \sum_{n=1}^{\infty} \mathrm{e}^{-\alpha^2 n^2} \cos n\xi\Big] = \frac{\mathrm{e}^{-\xi^2/(4\alpha^2)}}{2\sqrt{\pi}\,\alpha} + \frac{1}{2\pi\sqrt{\pi}\,\alpha}\int_\pi^\infty \mathrm{e}^{-x^2/(4\alpha^2)}\,\mathrm{d}x$$
$$+ \frac{1}{\pi\sqrt{\pi}\,\alpha}\sum_{n=1}^{\infty}\int_\pi^\infty \mathrm{e}^{-x^2/(4\alpha^2)} \cos nx \cos n\xi \,\mathrm{d}x. \tag{1.3.22}$$

By (1.3.20), we have

$$\varphi_\alpha(x) = \int_{-\pi}^{\pi} W_\alpha(x-\xi)\varphi(\xi)\,\mathrm{d}\xi = \int_{-\pi}^{\pi} W_\alpha(\xi)\varphi(x-\xi)\,\mathrm{d}\xi, \tag{1.3.23}$$

where

$$W_\alpha(x-\xi) = \frac{1}{\pi}\Big[\frac{1}{2} + \sum_{n=1}^{\infty} \mathrm{e}^{-\alpha^2 n^2} \cos n(x-\xi)\Big].$$

Multiplying both sides of the identity

$$1 = \int_{-\pi}^{\pi} W_\alpha(\xi)\,\mathrm{d}\xi$$

by $\varphi(x)$ and subtracting it from (1.3.23), we find

$$|\varphi(x) - \varphi_\alpha(x)| \le \int_{-\pi}^{\pi} |\varphi(x) - \varphi(x-\xi)|\cdot|W_\alpha(\xi)|\,\mathrm{d}\xi.$$

However, we have

$$|\varphi(x) - \varphi_\alpha(x-\xi)| \le \omega(|\xi|).$$

If $\mu > 0$, then, by the property of the continuity modulus, we have

$$\omega(|\xi|) = \omega\Big(\frac{1}{\mu}\mu|\xi|\Big) \le \omega\Big(\frac{1}{\mu}\Big)(\mu|\xi| + 1).$$

Thus, we obtain

$$|\varphi(x) - \varphi_\alpha(x)| \le \omega\Big(\frac{1}{\mu}\Big) \int_{-\pi}^{\pi} (\mu|\xi| + 1)|W_\alpha(\xi)|\,\mathrm{d}\xi$$

$$= \omega\Big(\frac{1}{\mu}\Big)\Big[\mu \int_{-\pi}^{\pi} |\xi|\cdot|W_\alpha(\xi)|\,\mathrm{d}\xi + \int_{-\pi}^{\pi} |W_\alpha(\xi)|\,\mathrm{d}\xi\Big] = \omega\Big(\frac{1}{\mu}\Big)[\mu I_1 + I_2],$$

where

$$I_1 = \int_{-\pi}^{\pi} |\xi|\cdot|W_\alpha(\xi)|\,\mathrm{d}\xi, \qquad I_2 = \int_{-\pi}^{\pi} |W_\alpha(\xi)|\,\mathrm{d}\xi.$$

Taking into account equality (1.3.22), we estimate I_1 and I_2:

$$I_1 \le \frac{2\alpha}{\sqrt{\pi}} + \Big[\frac{\pi}{4} + \frac{\pi^3}{24\alpha}\Big(\frac{2\sqrt{\pi}}{\alpha^2} + \frac{1}{\alpha} + 1\Big) - \frac{2\alpha}{\sqrt{\pi}}\Big]\,\mathrm{e}^{-\pi^2/(4\alpha^2)},$$

$$I_2 \le 1 + \frac{\pi^2}{6\sqrt{\pi}}\Big(1 + \frac{\pi}{2} + \frac{3\sqrt{\pi}}{2}\alpha\Big)\frac{\mathrm{e}^{-\pi^2/(4\alpha^2)}}{\alpha^3}.$$

Therefore, we have

$$|\varphi(x) - \varphi_\alpha(x)| \le \omega\Big(\frac{1}{\mu}\Big)\Big[1 + \frac{2\alpha\mu}{\sqrt{\pi}} + K(\alpha)\Big],$$

where

$$K(\alpha) = \Big[\frac{\pi}{4\alpha} + \frac{\pi^3}{24\alpha^2}\Big(\frac{2\sqrt{\pi}}{\alpha^2} + \frac{1}{\alpha} + 1\Big) - \frac{2}{\sqrt{\pi}} + \frac{\pi^2}{6\sqrt{\pi}\,\alpha^3}\Big(1 + \frac{\pi}{2} + \frac{3\sqrt{\pi}}{2}\alpha\Big)\Big]\,\mathrm{e}^{-\pi^2/(4\alpha^2)}.$$

Setting $\mu = 1/\alpha$, we obtain

$$|\varphi(x) - \varphi_\alpha(x)| \le \omega(\alpha)\Big[1 + \frac{2}{\sqrt{\pi}} + K(\alpha)\Big].$$

The theorem is proved. □

Corollary 1.3.1. For each $y \in [0, b]$, we have

$$\lim_{\alpha \to 0} u_{2\alpha}(x, y) = u_2(x, y)$$

uniformly relative to x in $[-\pi, \pi]$.

Assume that instead of $\varphi_j(x)$, $j = 1, 2$, we know their approximations $\varphi_{j\delta}(x)$ which are continuous, bounded and

$$|\varphi_j(x) - \varphi_{j\delta}(x)| \le \delta, \qquad j = 1, 2. \tag{1.3.24}$$

If we apply (1.3.20) to $\varphi_j(x)$, then instead of $u_{2\alpha}(x, y)$ we obtain its approximation

$$u_{2\alpha\delta}(x, y) = \frac{1}{4\pi^2} \int_{-\pi}^{\pi} \varphi_{3\delta}(\tau) R_\alpha(x - \tau, y)\,\mathrm{d}\tau. \tag{1.3.25}$$

Respectively, instead of $\varphi(x)$ we shall have $\varphi_{\alpha\delta}(x)$.

The function $\varphi_{\alpha\delta}(x)$ is an approximate solution to the problem. It can be found actually. Its deviation from the exact solution $\varphi(x)$ can be estimated by the following scheme (Lavrent'ev, 1973):

$$\begin{aligned} |\varphi(x) - \varphi_{\alpha\delta}(x)| &\le |\varphi(x) - \varphi_\alpha(x)| + |\varphi_\alpha(x) - \varphi_{\alpha\delta}(x)| \\ &\le \varepsilon_1(\alpha) + \varepsilon_2(\alpha, \delta). \end{aligned} \tag{1.3.26}$$

The estimate of the first term $\varepsilon_1(\alpha)$ is given by formula (1.3.21). To estimate $\varepsilon_2(\alpha, \delta)$ we use (1.3.20) setting $y = b$

$$\begin{aligned} |\varphi_\alpha(x) - \varphi_{\alpha\delta}(x)| &\le \frac{1}{4\pi^2} \int_{-\pi}^{\pi} |R_\alpha(x - \tau, b)| \cdot |\varphi_3(\tau) - \varphi_{3\delta}(\tau)|\,\mathrm{d}\tau \\ &\le \frac{\delta}{4\pi^2} \int_{-\pi}^{\pi} |R_\alpha(x - \tau, b)|\,\mathrm{d}\tau, \end{aligned} \tag{1.3.27}$$

$$\begin{aligned} |R_\alpha(x, b)| &\le \left| \int_{-\pi}^{\pi} \sum_{n=-\infty}^{\infty} \mathrm{e}^{-\alpha^2 n^2 - n(y-b) + \mathrm{i}n(x-\xi)}\,\mathrm{d}\xi \right| \\ &= 2\left| \int_{-\pi}^{\pi} \sum_{n=-\infty}^{\infty} \mathrm{e}^{-\alpha^2 n^2 - n(y-b)} \cos n(x - \xi)\,\mathrm{d}\xi \right| \\ &\le 2\pi \mathrm{e}^{b^2/(2\alpha^2)} \sum_{n=0}^{\infty} \mathrm{e}^{-\alpha^2 n^2/2}. \end{aligned} \tag{1.3.28}$$

Using equality (1.3.22), we obtain

$$\sum_{n=0}^{\infty} \mathrm{e}^{-(\alpha^2 n^2)/2} = \sqrt{\frac{\pi}{2}}\frac{1}{\alpha} + \frac{1}{\sqrt{2\pi}}\frac{1}{\alpha}\int_{\pi}^{\infty} \mathrm{e}^{-x^2/(4\alpha^2)}\,\mathrm{d}x + \sqrt{\frac{\pi}{2}}\frac{1}{\alpha}\sum_{n=1}^{\infty}\int_{\pi}^{\infty} \mathrm{e}^{-x^2/(4\alpha^2)}\cos nx\,\mathrm{d}x.$$

Since

$$\Big|\int_{\pi}^{\infty} \mathrm{e}^{-x^2/(4\alpha^2)}\cos nx\,\mathrm{d}x\Big| \le \frac{\pi \mathrm{e}^{-\pi^2/(2\alpha^2)}}{\alpha^2 n^2} + \frac{1}{\alpha^2 n^2}\int_{\pi}^{\infty} \mathrm{e}^{-x^2/(2\alpha^2)}\,\mathrm{d}x + \frac{1}{2\alpha^4 n^2}\int_{\pi}^{\infty} x^2 \mathrm{e}^{-x^2/(2\alpha^2)}\,\mathrm{d}x$$

and

1) $\mathrm{e}^{-\pi^2/(2\alpha^2)} \le 1,$ **2)** $\displaystyle\int_{\pi}^{\infty} \mathrm{e}^{-x^2/(2\alpha^2)}\,\mathrm{d}x \le \sqrt{2}\,\alpha,$

3) $\displaystyle\int_{\pi}^{\infty} x^2 \mathrm{e}^{-x^2/(2\alpha^2)}\,\mathrm{d}x \le (\pi + \sqrt{2}\,\alpha)\alpha^2,$ **4)** $\displaystyle\sum_{n=1}^{\infty}\frac{1}{n^2} = \frac{\pi^2}{6},$

therefore,

$$\sum_{n=0}^{\infty} \mathrm{e}^{-(\alpha^2 n^2)/2} \le \sqrt{\frac{\pi}{2}}\frac{1}{\alpha} + \frac{1}{\sqrt{\pi}} + \sqrt{\frac{2}{\pi}}\frac{\pi^2}{4}\Big(\frac{\pi}{\alpha^2} + \frac{\sqrt{2}}{\alpha^2}\Big) \le \frac{1}{2} + \frac{1}{\alpha} + \frac{12}{\alpha^2} + \frac{10}{\alpha^3}.$$

Therefore, from (1.3.28) we have

$$|R_\alpha(x,y)| \le 4\pi \mathrm{e}^{b^2/2\alpha^2}\Big(\frac{1}{2} + \frac{1}{\alpha} + \frac{12}{\alpha^2} + \frac{10}{\alpha^3}\Big).$$

Correspondingly, (1.3.27) takes the form

$$|\varphi_\alpha(x) - \varphi_{\alpha\delta}(x)| \le 2\mathrm{e}^{b^2/2\alpha^2}\Big(\frac{1}{2} + \frac{1}{\alpha} + \frac{12}{\alpha^2} + \frac{10}{\alpha^3}\Big)\delta. \tag{1.3.29}$$

Theorem 1.3.3. *Suppose that*

$$|\varphi_3(x) - \varphi_{3\delta}(x)| \le 2\delta.$$

If $\alpha = \alpha(\delta)$ *is the root of the equation*

$$\left(\frac{1}{2} + \frac{1}{\alpha} + \frac{12}{\alpha^2} + \frac{10}{\alpha^3}\right) e^{b^2/2\alpha^2} = \frac{1}{\sqrt{\delta}}, \tag{1.3.30}$$

and $u_{2\alpha\delta}(x, y)$ *is defined by formula* (1.3.25), *then*

$$\lim_{\delta \to 0} u_{2\alpha\delta}(x, y) = u_2(x, y)$$

uniformly in the whole half-strip.

Proof. It suffices to show that

$$\lim_{\delta \to 0} \varphi_{\alpha\delta}(x) = \varphi(x).$$

If (1.3.30) holds, then

$$\varepsilon_2(\alpha, \delta) = \sqrt{\delta}.$$

Therefore,

$$\lim_{\delta \to 0} \varepsilon_2(\alpha(\delta), \delta) = 0.$$

Besides, as follows from (1.3.30),

$$\lim_{\delta \to 0} \alpha(\delta) = 0.$$

Therefore, since the inverse function exists, we have

$$\lim_{\delta \to 0} \varepsilon_1(\alpha(\delta)) = 0.$$

Then (1.3.26) yields

$$\lim_{\delta \to 0} \varphi_{\alpha\delta}(x) = \varphi(x), \qquad \alpha = \alpha(\delta).$$

□

Theorem 1.3.3 allows us to approximate the solution with arbitrary accuracy if

$$\lim_{\delta \to 0} \varphi_{j\delta}(x) = \varphi_j(x), \quad j = 1, 2,$$

and if the exact solution exists and is continuous on the x-axis.

Evidently,

$$u_{\alpha(\delta)}(x, y) = u_{1\delta}(x, y) + y u_{2\alpha(\delta)}(x, y)$$

will be an approximate solution of the initial problem.

1.4. THE FIRST INTERNAL BOUNDARY VALUE PROBLEM IN THE HALF-PLANE

We need to find the solution of the biharmonic equation

$$\Delta^2 u(x,y) = 0 \tag{1.4.1}$$

in the domain $D_1 = \{y > 0\}$ under the conditions

$$\begin{gathered} u \to 0 \quad \text{for} \quad y \to \infty, \\ u(x,0) = \varphi_1(x), \qquad u(x,b) = \varphi_2(x). \end{gathered} \tag{1.4.2}$$

Here the functions φ_j, $j = 1, 2$, are given, and

$$\varphi_j(x) = O(|x|^{-2-\varepsilon}), \quad \varepsilon > 0.$$

This problem is ill-posed. Really, by the representation (Tikhonov and Samarskii, 1990)

$$u(x,y) = u_1(x,y) + y u_2(x,y), \tag{1.4.3}$$

which holds for each biharmonic function (u_j, $j = 1, 2$, are functions harmonic for $y > 0$), we obtain the following problems under conditions (1.4.2):

$$\Delta u_1(x,y) = 0, \quad y > 0, \qquad u_1(x,\infty) = 0, \quad u_1(x,0) = \varphi_1(x), \tag{1.4.4}$$

$$\Delta u_2(x,y) = 0, \quad y > 0, \qquad u_2(x,\infty) = 0, \quad u_2(x,b) = \varphi_3(x). \tag{1.4.5}$$

Here

$$u'_{2y} = \frac{1}{2}\,\Delta u(x,y), \qquad \varphi_3(x) = \frac{1}{b}[\varphi_2(x) - u_1(x,b)],$$

$$\varphi_j(x) = O(|x|^{-2-\varepsilon}), \quad j = 1, 2, 3, \quad \varepsilon > 0.$$

Hence we see that the problem (1.4.5) is an ill-posed problem of harmonic continuation from $D_2 = \{y > b\}$ into the strip $0 < y < b$ (Lavrent'ev, 1973, p. 9). However, we can impose certain additional information on $u_2(x,y)$ which provides the conditional well-posedness of the problem (1.4.5) (Lavrent'ev, 1973, p. 42). Therefore, the problem (1.4.1), (1.4.2) will be conditionally well-posed. So we can construct approximations to the desired solution.

In this section we shall show that the condition of existence of a solution to the problem (1.4.5) gives us the possibility to approximate to the solution of (1.4.1), (1.4.2) arbitrarily close. If, besides, we know the modulus of continuity of the function $u_2(x,0)$, then we can estimate the error of the approximate solution of the problem (1.4.1), (1.4.2).

First, we assume that φ_j, $j = 1,2$, are known exactly. We shall solve the problems (1.4.4) and (1.4.5) by applying the Fourier transform with respect to x. We have

$$u_1(x,y) = \frac{1}{2\pi}\int_{-\infty}^{\infty} \widehat{\varphi}_1(\omega)\mathrm{e}^{-|\omega|y-\mathrm{i}\omega x}\,\mathrm{d}\omega, \tag{1.4.6}$$

$$u_2(x,y) = \frac{1}{2\pi}\int_{-\infty}^{\infty} \widehat{\varphi}_3(\omega)\mathrm{e}^{-|\omega|(y-b)-\mathrm{i}\omega x}\,\mathrm{d}\omega, \tag{1.4.7}$$

where

$$\widehat{\varphi}_j(\omega) = \int_{-\infty}^{\infty} \varphi_j(x)\mathrm{e}^{-\mathrm{i}\omega x}\,\mathrm{d}x, \qquad j = 1,3.$$

Note that for the values $y \in [0,b]$ formula (1.4.7) is unstable. Therefore, regularizing (1.4.7) similarly as in (Ivanov, 1965), we obtain

$$u_{2\alpha}(x,y) = \frac{1}{2\pi}\int_{-\infty}^{\infty} R_\alpha(x-t,y)\varphi_3(t)\,\mathrm{d}t, \tag{1.4.8}$$

where

$$R_\alpha(x,y) = \frac{1}{2\pi}\int_{-\infty}^{\infty} \mathrm{e}^{-\alpha^2\omega^2-|\omega|(b-y)}\cos\omega x\,\mathrm{d}\omega$$

and α is the regularization parameter. Calculating the derivative of $R_\alpha(x,y)$ with respect to x and integrating it for $y = 0$ by parts, we get

$$R'_{\alpha x}(x,0) = -\frac{x}{2\alpha^2}R_\alpha(x,0) - \frac{b}{2\alpha^2}\int_0^{\infty} \mathrm{e}^{-\alpha^2\omega^2+\omega b}\sin\omega x\,\mathrm{d}\omega.$$

Since

$$R_\alpha|_{x=0} = \frac{1}{2\alpha\sqrt{\pi}}\mathrm{e}^{b^2/(4\alpha^2)} - \frac{1}{\alpha\pi}\mathrm{e}^{b^2/(4\alpha^2)}\int_0^{b/(2\alpha)} \mathrm{e}^{-t^2}\,\mathrm{d}t,$$

therefore, solving the differential equation of first order for $R_\alpha(x,0)$ by the method of variation of constants, we obtain

$$\begin{aligned} R_\alpha(x,0) = &-\frac{b}{2\alpha^2\pi}\mathrm{e}^{x^2/(4\alpha^2)}\int_0^{\infty} \mathrm{e}^{-\alpha^2\omega^2+\omega b}\Big(\int_0^{x} \mathrm{e}^{\tau^2/(4\alpha^2)}\sin\omega\tau\,\mathrm{d}\tau\Big)\,\mathrm{d}\omega \\ &+ \Big[\frac{1}{2\alpha\sqrt{\pi}} - \frac{1}{\alpha\sqrt{\pi}}\int_0^{b/(2\alpha)} \mathrm{e}^{-t^2}\,\mathrm{d}t\Big]\mathrm{e}^{-(x^2-b^2)/(4\alpha^2)}. \end{aligned}$$

Hence, taking into account the strict positiveness of the first and last terms, we obtain the estimate

$$\int_{-\infty}^{\infty} |R_\alpha(x,0)|\,\mathrm{d}x \le \Big(1+\frac{b}{\alpha\sqrt{\pi}}\Big)\mathrm{e}^{b^2/(4\alpha^2)}. \tag{1.4.9}$$

The function $u_{2\alpha}(x,y)$ defined by (1.4.8) may be considered as an approximate solution of the problem (1.4.5).

Let

$$u_2(x,0)=\psi(x), \qquad u_{2\alpha}(x,0)=\psi_\alpha(x).$$

Under the conditions stated above, the function $\psi(x)$ exists and is unique. Let $\omega(h)$ be the continuity modulus of $\psi(x)$, then

$$|\psi(x)-\psi_\alpha(x)| \le c\omega(\alpha), \tag{1.4.10}$$

where $c = \text{const} > 0$ is independent of α.

Hence it follows that

$$\lim_{\alpha\to 0}\psi_\alpha(x)=\psi(x)$$

uniformly relative to x in the whole axis $(-\infty,\infty)$.

Analogously we prove for each $y\in[0,\infty]$ the equality

$$\lim_{\alpha\to 0} u_{2\alpha}(x,y)=u(x,y) \tag{1.4.11}$$

uniformly relative to x in $(-\infty,\infty)$.

We assume now that instead of $\varphi_j(x)$, $j=1,2$, we are given their approximations $\varphi_{j\delta}(x)$ such that

$$|\varphi_j(x)-\varphi_{j\delta}(x)| \le \delta. \tag{1.4.12}$$

If we apply (1.4.8) to $\varphi_{j\delta}(x)$, then, instead of $u_{2\alpha}(x,y)$, we obtain its approximation

$$u_{2\alpha\delta}(x,y)=\frac{1}{2\pi}\int_{-\infty}^{\infty} R_\alpha(x-t,y)\varphi_{3\delta}(t)\,\mathrm{d}t. \tag{1.4.13}$$

Accordingly, instead of $\psi_\alpha(x)$ we shall have $\psi_{\alpha\delta}(x)$. Estimate the deviation of $\psi_{\alpha\delta}(x)$ from $\psi(x)$

$$|\psi(x)-\psi_{\alpha\delta}(x)| \le |\psi(x)-\psi_\alpha(x)|+|\psi_\alpha(x)-\psi_{\alpha\delta}(x)| \le \varepsilon_1(\alpha)+\varepsilon_2(\alpha,\delta). \tag{1.4.14}$$

The estimate of the first term $\varepsilon_1(\alpha)$ is given by formula (1.4.10). To estimate $\varepsilon_2(\alpha,\delta)$ we use (1.4.8) and (1.4.12) setting $y=0$. Then

$$\begin{aligned}|\psi_\alpha(x)-\psi_{\alpha\delta}(x)| &\le \frac{1}{2\pi}\int_{-\infty}^{\infty}|R_\alpha(x-t,0)|\cdot|\varphi_3(x)-\varphi_{3\delta}(x)|\,\mathrm{d}t\\ &\le \frac{\delta}{b\pi}\int_{-\infty}^{\infty}|R_\alpha(x-t,0)|\,\mathrm{d}t,\end{aligned}\tag{1.4.15}$$

because of

$$\varphi_{3\delta}(x)=\frac{1}{b}[\varphi_{2\delta}(x)-u_{1\delta}(x,b)],\qquad |\varphi_3(t)-\varphi_{3\delta}(t)|\le\frac{2\delta}{b}.$$

Estimating the integral by means of (1.4.9), we find

$$|\psi_\alpha(x)-\psi_{\alpha\delta}(x)|\le\Big(1+\frac{b}{\alpha\sqrt{\pi}}\Big)\frac{\delta \mathrm{e}^{b^2/(4\alpha^2)}}{b}.\tag{1.4.16}$$

Therefore,

$$\lim_{\delta\to 0}\psi_{\alpha\delta}(x)=\psi(x)$$

and

$$\lim_{\delta\to 0}u_{2\alpha\delta}(x,y)=u(x,y)$$

uniformly in the whole half-plane. Finally, we obtain the approximate solution of the problem (1.4.1), (1.4.2)

$$u_{\alpha(\delta)\delta}(x,y)=u_{1\delta}(x,y)+yu_{2\alpha(\delta)\delta}(x,y),\tag{1.4.17}$$

where $\alpha=\alpha(\delta)$ is the root of the equation

$$\Big(1+\frac{b}{\alpha\sqrt{\pi}}\Big)\mathrm{e}^{b^2/(4\alpha^2)}=\frac{b}{\sqrt{\pi}}.\tag{1.4.18}$$

Evidently,

$$\lim_{\delta\to 0}u_{\alpha(\delta)\delta}(x,y)=u(x,y)$$

uniformly in $\{y>0\}$.

The approximate solution of the problem may be taken as the integral with finite limits

$$u_{\alpha(\delta)N(\delta)}(x,y)=\frac{y}{\pi}\int_{-N(\delta)}^{N(\delta)}\frac{\varphi_{1\delta}(t)\,\mathrm{d}t}{(x-t)^2+y^2}+\frac{y}{2\pi}\int_{-N(\delta)}^{N(\delta)}R_\alpha(x-t,y)\varphi_{3\delta}(t)\,\mathrm{d}t.\tag{1.4.19}$$

The deviation from the exact solution on the x-axis can be estimated as follows

$$\begin{aligned}
&\left|u_{\alpha(\delta)N(\delta)}(x,0) - u(x,0)\right| \\
&\quad \le \left|u_{1N(\delta)}(x,0) - u_1(x,0)\right| + \left|y[u_{2\alpha(\delta)}(x,y) - u_2(x,y)]_{y=0}\right| \\
&\quad\quad + \left|y[u_{2\alpha(\delta)N(\delta)}(x,y) - u_{2\alpha(\delta)}(x,y)]_{y=0}\right| \\
&\quad \le \varepsilon_1(N(\delta)) + \varepsilon_2(\alpha(\delta)) + \varepsilon_3(\alpha(\delta), N(\delta)). \qquad (1.4.20)
\end{aligned}$$

Here

$$\varepsilon_1(N(\delta)) \le \frac{C_1+\delta}{\pi}\Big(\pi - \arctan\frac{1}{N(\delta)}\Big),$$

$$C_1 = \max|\varphi_1(x)| \ll \varepsilon_2(\alpha(\delta)) = C\omega(\alpha(\delta)), \qquad C = \text{const} > 0,$$

$$\varepsilon_3(\alpha(\delta), N(\delta)) = \frac{C_3+\delta}{\pi^{3/2}}\exp\Big\{\frac{b^2 - N^2(\delta)}{4\alpha^2(\delta)}\Big\},$$

$$C_3 = C_1 + \max|\varphi_2(x)|.$$

The parameters $\alpha(\delta)$ and $N(\delta)$ are found from equations (1.4.18) and

$$\frac{C_1+\delta}{\pi}\Big(\pi - \arctan\frac{1}{N}\Big) + \frac{C_3+\delta}{\pi^{3/2}}\,\mathrm{e}^{(b^2-N^2)/(4\alpha^2)} = \sqrt{\delta}, \qquad (1.4.21)$$

respectively.

Remark 1.4.1. Suppose that the last condition in (1.4.2) is given on the curve

$$y = f(x) \in C'(-\infty,\infty), \qquad |f(\pm\infty)| < \infty, \qquad b = \max f(x).$$

Then, solving the Dirichlet problem (1.4.5) for the Laplace equation in the infinite domain $\{y > f(x)\}$ and taking the values u_j, $j = 1, 2$, for $y = b$, by representation (1.4.3), we derive the values of u on the straight line $y = b$.

1.5. THE FIRST INTERNAL BOUNDARY VALUE PROBLEM IN RECTANGLES

Let D_j, $j = 1, 2$, be the rectangles

$$D_1 = \{(x,y) \mid 0 < x < a_3,\ 0 < y < b_3\},$$

$$D_2 = \{(x, y) \mid a_1 < x < a_2,\ b_1 < y < b_2\}$$

bounded by

$$S_1 = \partial D_1 = \{(x, y) \mid x = 0, a_3;\ y = 0, b_3\},$$
$$S_2 = \partial D_2 = \{(x, y) \mid x = a_1, a_2;\ y = b_1, b_2\}.$$

We need to find the function $u(x, y)$ in D_1 satisfying the conditions

$$\Delta^2 u(x, y) = 0 \qquad \text{in} \quad D_1, \tag{1.5.1}$$
$$u(0, y) = \psi_1(y), \qquad u(a_3, y) = \psi_2(y), \qquad 0 \le y \le b_3, \tag{1.5.2}$$
$$u(x, 0) = \varphi_1(x), \qquad u(x, b_3) = \varphi_2(x), \qquad 0 \le x \le a_3, \tag{1.5.3}$$
$$u(a_1, y) = \psi_3(y), \qquad u(a_2, y) = \psi_4(y), \qquad b_1 \le y \le b_2, \tag{1.5.4}$$
$$u(x, b_1) = \varphi_3(x), \qquad u(x, b_2) = \varphi_4(x), \qquad a_1 \le x \le a_2. \tag{1.5.5}$$

Here Δ is the Laplace operator, $\psi_j(y)$, $\varphi_j(x)$ are given functions such that

$$\psi_3(b_1) = \varphi_3(a_1), \qquad \psi_4(b_1) = \varphi_3(a_2),$$
$$\psi_3(b_2) = \varphi_4(a_1), \qquad \psi_4(b_2) = \varphi_4(a_2).$$

We shall seek the solution of IBVP 1 (1.5.1)–(1.5.5) as the series

$$\begin{aligned} u(x, y) &= \sum_{k=1}^{\infty} \Big\{ \big[a_k^{(1)} A_k^{(1)}(y) + a_k^{(2)} A_k^{(2)}(y) + b_k^{(1)} B_k^{(1)}(y) + b_k^{(2)} B_k^{(2)}(y)\big] \sin \frac{k\pi x}{a_3} \\ &\quad + \big[a_k^{(3)} A_k^{(3)}(x) + a_k^{(4)} A_k^{(4)}(x) + b_k^{(3)} B_k^{(3)}(x) + b_k^{(4)} B_k^{(4)}(x)\big] \sin \frac{k\pi y}{b_3} \Big\}. \end{aligned} \tag{1.5.6}$$

Here $a_k^{(i)}$, $b_k^{(i)}$ $(k = \overline{1, \infty},\ i = \overline{1, 4})$ are arbitrary constants such that

$$|a_k^{(i)}|, |b_k^{(i)}| < C, \tag{1.5.7}$$

where C is a nonnegative number;

$$A_k^{(1)}(y) = \frac{1}{\sinh \frac{k\pi b_3}{a_3}} \sinh \frac{k\pi y}{a_3}, \qquad A_k^{(2)}(y) = \frac{1}{\sinh \frac{k\pi b_3}{a_3}} \sinh \frac{k\pi (b_3 - y)}{a_3},$$
$$A_k^{(3)}(x) = \frac{1}{\sinh \frac{k\pi a_3}{b_3}} \sinh \frac{k\pi x}{b_3}, \qquad A_k^{(4)}(x) = \frac{1}{\sinh \frac{k\pi a_3}{b_3}} \sinh \frac{k\pi (a_3 - x)}{b_3},$$

$$B_k^{(1)}(y) = \frac{a_3}{2k\pi \sinh^2 \dfrac{k\pi b_3}{a_3}} \left[b_3 \sinh \frac{k\pi y}{a_3} - y \sinh \frac{k\pi b_3}{a_3} \cosh \frac{k\pi(b_3 - y)}{a_3} \right],$$

$$B_k^{(2)}(y) = \frac{a_3}{2k\pi \sinh^2 \dfrac{k\pi b_3}{a_3}} \left[y \sinh \frac{k\pi b_3}{a_3} \cosh \frac{k\pi y}{a_3} - b_3 \cosh \frac{k\pi b_3}{a_3} \sinh \frac{k\pi y}{a_3} \right],$$

$$B_k^{(3)}(x) = \frac{b_3}{2k\pi \sinh^2 \dfrac{k\pi a_3}{b_3}} \left[a_3 \sinh \frac{k\pi x}{b_3} - x \sinh \frac{k\pi a_3}{b_3} \cosh \frac{k\pi(a_3 - x)}{b_3} \right],$$

$$B_k^{(4)}(x) = \frac{b_3}{2k\pi \sinh^2 \dfrac{k\pi a_3}{b_3}} \left[x \sinh \frac{k\pi a_3}{b_3} \cosh \frac{k\pi x}{b_3} - a_3 \cosh \frac{k\pi a_3}{b_3} \sinh \frac{k\pi x}{b_3} \right].$$

Theorem 1.5.1. *If* (1.5.7) *holds, then the solution of IBVP 1 of the form* (1.5.6) *is unique; i. e., relations*

$$u(0, y) = u(a_3, y) = 0, \qquad 0 \le y \le b_3, \tag{1.5.8}$$

$$u(a_1, y) = u(a_2, y) = 0, \qquad b_1 \le y \le b_2, \tag{1.5.9}$$

$$u(x, 0) = u(x, b_3) = 0, \qquad 0 \le x \le a_3, \tag{1.5.10}$$

$$u(x, b_1) = u(x, b_2) = 0, \qquad a_1 \le x \le a_2, \tag{1.5.11}$$

imply $u(x, y) = 0$ *in* D_1.

Proof. Formulas (1.5.8), (1.5.9), and (1.5.6) yield

$$a_k^{(i)} = 0, \quad k = \overline{1, \infty}, \quad i = \overline{1, 4}.$$

Now formulas (1.5.10) and (1.5.11) imply that the functions

$$\psi_j(y) = \sum_{k=1}^{\infty} \Big\{ \big[b_k^{(1)} B_k^{(1)}(y) + b_k^{(2)} B_k^{(2)}(y) \big] \sin \frac{k\pi a_{j-2}}{a_3} + \big[b_k^{(3)} B_k^{(3)}(a_{j-2}) + b_k^{(4)} B_k^{(4)}(a_{j-2}) \big] \sin \frac{k\pi y}{b_3} \Big\}, \quad j = 3, 4,$$

vanish on the segment $[b_1, b_2]$, and the functions

$$\varphi_j(x) = \sum_{k=1}^{\infty} \Big\{ \big[b_k^{(1)} B_k^{(1)}(b_{j-2}) + b_k^{(2)} B_k^{(2)}(b_{j-2}) \big] \sin \frac{k\pi x}{a_3}$$

$$+ \left[b_k^{(3)} B_k^{(3)}(x) + b_k^{(4)} B_k^{(4)}(x)\right] \sin \frac{k\pi b_{j-2}}{b_3}\Big\}, \quad j = 3, 4,$$

vanish on the segment $[a_1, a_2]$.

As follows from (1.5.7), the functions $\psi_j(y)$ and $\varphi_j(x)$ $(j = 3, 4)$ admit analytic continuation into the rectangles

$$\overline{P} = \Big\{z = y + \mathrm{i}\xi \;\Big|\; -\frac{a_3 d}{\pi} \le \operatorname{Re} z \le \frac{a_3 d}{\pi}, \quad -\frac{b_3 d}{\pi} \le \operatorname{Im} z \le \frac{b_3 d}{\pi}\Big\},$$

$$\overline{\Pi} = \Big\{\zeta = x + \mathrm{i}\eta \;\Big|\; -\frac{b_3 d}{\pi} \le \operatorname{Re} \zeta \le \frac{b_3 d}{\pi}, \quad -\frac{a_3 d}{\pi} \le \operatorname{Im} \zeta \le \frac{a_3 d}{\pi}\Big\},$$

$$d > 2 \max\Big\{\frac{a_3}{b_3}, \frac{b_3}{a_3}\Big\}\pi,$$

by means of the equalities

$$\psi_j(z) = \sum_{k=1}^{\infty} \Big[a_{jk,0}^{[1]} \mathrm{e}^{(k\pi/a_3)z} + a_{jk,1}^{[1]} z \mathrm{e}^{(k\pi/a_3)z} + a_{jk,0}^{[2]} \mathrm{e}^{-(k\pi/a_3)z}$$
$$+ a_{jk,1}^{[2]} z \mathrm{e}^{-(k\pi/a_3)z} + a_{jk,0}^{[3]} \mathrm{e}^{\mathrm{i}(k\pi/b_3)z} + a_{jk,0}^{[4]} \mathrm{e}^{-\mathrm{i}(k\pi/b_3)z}\Big],$$

$$\varphi_j(\zeta) = \sum_{k=1}^{\infty} \Big[b_{jk,0}^{[1]} \mathrm{e}^{(k\pi/b_3)\zeta} + b_{jk,1}^{[1]} \zeta \mathrm{e}^{(k\pi/b_3)\zeta} + b_{jk,0}^{[2]} \mathrm{e}^{-(k\pi/b_3)\zeta}$$
$$+ b_{jk,1}^{[2]} \zeta \mathrm{e}^{-(k\pi/b_3)\zeta} + b_{jk,0}^{[3]} \mathrm{e}^{\mathrm{i}(k\pi/a_3)\zeta} + b_{jk,0}^{[4]} \mathrm{e}^{-\mathrm{i}(k\pi/a_3)\zeta}\Big]. \quad (1.5.12)$$

Here

$$a_{jk,0}^{[1]} = H_{1k} \sin \frac{k\pi a_{j-2}}{a_3} \Big[b_k^{(1)} b_3 - b_k^{(2)} b_3 \cosh \frac{k\pi b_3}{a_3}\Big],$$
$$a_{jk,1}^{[1]} = H_{1k} \sin \frac{k\pi a_{j-2}}{a_3} \Big[-b_k^{(1)} \mathrm{e}^{-(k\pi/a_3)b_3} \sinh \frac{k\pi b_3}{a_3} + b_k^{(2)} \sinh \frac{k\pi b_3}{a_3}\Big],$$
$$a_{jk,0}^{[2]} = H_{1k} \sin \frac{k\pi a_{j-2}}{a_3} \Big[-b_k^{(1)} b_3 - b_k^{(2)} b_3 \cosh \frac{k\pi b_3}{a_3}\Big],$$
$$a_{jk,1}^{[2]} = H_{1k} \sin \frac{k\pi a_{j-2}}{a_3} \Big[-b_k^{(1)} \mathrm{e}^{-(k\pi/a_3)b_3} \sinh \frac{k\pi b_3}{a_3} + b_k^{(2)} \sinh \frac{k\pi b_3}{a_3}\Big],$$
$$a_{jk,0}^{[3]} = -\frac{\mathrm{i}}{2} \left[b_k^{(3)} B_k^{(3)}(a_{j-2}) + b_k^{(4)} B_k^{(4)}(a_{j-2})\right],$$
$$a_{jk,0}^{[4]} = \frac{\mathrm{i}}{2} \left[b_k^{(3)} B_k^{(3)}(a_{j-2}) + b_k^{(4)} B_k^{(4)}(a_{j-2})\right],$$
$$b_{jk,0}^{[1]} = H_{2k} \sin \frac{k\pi b_{j-2}}{b_3} \Big[b_k^{(3)} a_3 - b_k^{(4)} a_3 \cosh \frac{k\pi a_3}{b_3}\Big],$$

$$b^{[1]}_{jk,1} = H_{2k} \sin \frac{k\pi b_{j-2}}{b_3} \left[-b^{(3)}_k \mathrm{e}^{-(k\pi/b_3)a_3} \sinh \frac{k\pi a_3}{b_3} + b^{(4)}_k \sinh \frac{k\pi a_3}{b_3} \right],$$
$$b^{[2]}_{jk,0} = H_{2k} \sin \frac{k\pi b_{j-2}}{b_3} \left[-b^{(3)}_k a_3 + b^{(4)}_k a_3 \cosh \frac{k\pi a_3}{b_3} \right],$$
$$b^{[2]}_{jk,1} = H_{2k} \sin \frac{k\pi b_{j-2}}{b_3} \left[-b^{(3)}_k \mathrm{e}^{(k\pi/b_3)a_3} \sinh \frac{k\pi a_3}{b_3} + b^{(4)}_k \sinh \frac{k\pi a_3}{b_3} \right],$$
$$b^{[3]}_{jk,0} = -\frac{\mathrm{i}}{2} \left[b^{(1)}_k B^{(1)}_k (b_{j-2}) + b^{(2)}_k B^{(2)}_k (b_{j-2}) \right],$$
$$b^{[4]}_{jk,0} = \frac{\mathrm{i}}{2} \left[b^{(1)}_k B^{(1)}_k (b_{j-2}) + b^{(2)}_k B^{(2)}_k (b_{j-2}) \right], \qquad j = 3, 4,$$

$$H_{1k} = \frac{a_3}{4k\pi \sinh^2 \dfrac{k\pi b_3}{a_3}}, \qquad H_{2k} = \frac{b_3}{4k\pi \sinh^2 \dfrac{k\pi a_3}{b_3}}.$$

Since $\psi_j(z) = 0$, $j = 3, 4$, on the real segment $[b_1, b_2]$ and $\varphi_j(\zeta) = 0$, $j = 3, 4$, on the real segment $[a_1, a_2]$, therefore, taking into account the uniqueness of continuation of analytic functions of one complex variable, we have

$$\psi_j(z) = 0, \quad j = 3, 4, \quad z \in \overline{P},$$
$$\varphi_j(\zeta) = 0, \quad j = 3, 4, \quad \zeta \in \overline{\Pi}.$$

We introduce the function

$$Q_1(\lambda) = \lambda^{-3} \sinh \lambda b_3 \sin^2 \lambda a_3.$$

The function $Q_1(\lambda)$ is an entire function of exponential type, which at the points

$$\lambda^{[1]}_k = \frac{k\pi}{a_3}, \quad \lambda^{[2]}_k = -\frac{k\pi}{a_3}, \quad \lambda^{[3]}_k = \mathrm{i}\frac{k\pi}{b_3}, \quad \lambda^{[4]}_k = -\mathrm{i}\frac{k\pi}{b_3}, \quad k = \overline{1, \infty},$$

has zeros of the following orders:

$$\nu^{[1]}_k = 2, \quad \nu^{[2]}_k = 2, \quad \nu^{[3]}_k = 1, \quad \nu^{[4]}_k = 1, \quad k = \overline{1, \infty}.$$

We will denote by $\overline{P_1}$ the conjugate diagram of the function $Q_1(\lambda)$. Let us clarify the form of the domain $\overline{P_1}$ in our case. The functions $Q_1(\lambda)$ and

$$\begin{aligned} L(\lambda) &= \sinh \lambda b_3 \sin^2 \lambda a_3 \\ &= \frac{1}{4}\mathrm{e}^{b_3\lambda} - \frac{1}{4}\mathrm{e}^{-b_3\lambda} - \frac{1}{8}\mathrm{e}^{(b_3+2a_3\mathrm{i})\lambda} - \frac{1}{8}\mathrm{e}^{(b_3-2a_3\mathrm{i})\lambda} \\ &\quad + \frac{1}{8}\mathrm{e}^{(-b_3+2a_3\mathrm{i})\lambda} + \frac{1}{8}\mathrm{e}^{(-b_3-2a_3\mathrm{i})\lambda} \end{aligned}$$

have the same indicatrix of growth; therefore, they have the same conjugate diagrams. The Borel functional transform of $L(\lambda)$ is as follows

$$\gamma(z) = \frac{1}{4}\cdot\frac{1}{z-b_3} - \frac{1}{4}\cdot\frac{1}{z+b_3} - \frac{1}{8}\cdot\frac{1}{z-(b_3+2a_3\mathrm{i})} - \frac{1}{8}\cdot\frac{1}{z-(b_3-2a_3\mathrm{i})} + \frac{1}{8}\cdot\frac{1}{z-(-b_3+2a_3\mathrm{i})} + \frac{1}{8}\cdot\frac{1}{z-(-b_3-2a_3\mathrm{i})}$$

and it has singularities only at the points

$$z = b_3,\ -b_3,\ b_3+2a_3\mathrm{i},\ b_3-2a_3\mathrm{i},\ -b_3+2a_3\mathrm{i},\ -b_3-2a_3\mathrm{i}.$$

Therefore,

$$\overline{P_1} = \{-b_3 \le \operatorname{Re} z \le b_3, \quad -2a_3 \le \operatorname{Im} z \le 2a_3\}.$$

By the hypotheses of Theorem 1.5.1, we have $\overline{P_1} \subset P$. Suppose now that C_1 is a simple closed contour lying in $P \setminus \overline{P_1}$. Let us denote by $A^{[i]}_{k,m}(t)$ the Borel functional transform of the entire function

$$\frac{Q_1(\lambda)}{(\lambda-\lambda^{[i]}_k)^{m+1}}, \qquad m = 0,1,2,\dots,\nu^{[i]}_k - 1, \qquad i = 1,2,3,4.$$

We set

$$M^{[i]}_{1k}(\mu) = (\mu-\lambda^{[i]}_k)^{\nu^{[i]}_k} Q_1^{-1}(\mu),$$

$$\psi^{[i]}_{k,m}(t) = \frac{1}{m!\,(\nu^{[i]}_k - m - 1)!} \times \sum_{q=0}^{\nu^{[i]}_k - m - 1} q!\cdot C^q_{\nu^{[i]}_k - m - 1} \left[M^{[i]}_{1k}(\mu)\right]^{(\nu^{[i]}_k - m - q - 1)}_{\mu=\lambda^{[i]}_k} A^{[i]}_{k,q}(t).$$

By direct computation we can show that

$$\frac{1}{2\pi\mathrm{i}}\int_{C_1} t^s e^{\lambda^{[i]}_p t}\psi^{[n]}_{k,m}(t)\,\mathrm{d}t = \begin{cases} 1, & i=n,\ p=k,\ s=m; \\ 0, & \text{in other cases.}\end{cases} \tag{1.5.13}$$

Multiply both sides of equalities (1.5.12) by the functions

$$\frac{1}{2\pi\mathrm{i}}\psi^{[3]}_{k,0}(z), \qquad k = \overline{1,\infty},$$

and integrate around the contour C_1. Then, taking into account (1.5.13), we obtain

$$a_{jk,0}^{[3]} = \frac{1}{2\pi i} \int_{C_1} \psi_j(z)\psi_{k,0}^{[3]}(z)\,dz, \qquad j = 3, 4, \quad k = \overline{1, \infty}. \qquad (1.5.14)$$

Analogously, we obtain

$$b_{jk,0}^{[3]} = \frac{1}{2\pi i} \int_{C_2} \varphi_j(\zeta)\varphi_{k,0}^{[3]}(\zeta)\,d\zeta, \qquad j = 3, 4, \quad k = \overline{1, \infty}. \qquad (1.5.15)$$

Here C_2 is a simple closed contour lying in $\Pi \setminus \overline{\Pi_1}$ where

$$\overline{\Pi_1} = \{-a_3 \le \operatorname{Re}\zeta \le a_3, \ -2b_3 \le \operatorname{Im}\zeta \le 2b_3\}, \qquad \overline{\Pi_1} \subset \Pi.$$

Recall that

$$\psi_j(z) = 0, \quad j = 3, 4, \quad z \in P,$$
$$\varphi_j(\zeta) = 0, \quad j = 3, 4, \quad \zeta \in \Pi.$$

Therefore, (1.5.14) and (1.5.15) yield

$$a_{jk,0}^{[3]} = 0, \quad b_{jk,0}^{[3]} = 0, \quad j = 3, 4, \quad k = \overline{1, \infty}.$$

Hence it follows that $b_k^{(1)} = b_k^{(2)} = b_k^{(3)} = b_k^{(4)} = 0$ for all $k = 1, 2, \ldots$.

Returning to (1.5.6) we conclude that $u(x, y) = 0$ in D_1. Note that the proof of the theorem also follows from the Jensen inequality for entire functions. □

1.6. THE FIRST INTERNAL BOUNDARY VALUE PROBLEM FOR STAR-LIKE DOMAINS

Let $r = R_j(q)$, $j = 1, 2$, be the equations for Lyapunov contours S_j, $j = 1, 2$, of star-like domains D_j, $j = 1, 2$; and let $G(p, q)$ be the Green function of the Dirichlet problem for the Laplace equation in D_1. We shall use the representation

$$u(p) = u_1(p) + r^2 u_2(p) \qquad (1.6.1)$$

for a biharmonic function $u(p)$ in the star-like domain D_1. Here

$$u_j(p) = \int_{S_1} \mu_j(q) \frac{\partial G(p, q)}{\partial n_q}\, d_q S_1, \qquad j = 1, 2, \qquad (1.6.2)$$

are harmonic functions. Taking into account condition (2), we obtain the integral Fredholm equation of first kind

$$\int_{S_1} \mu_2(q) K(p,q)\, \mathrm{d}_q S_1 = f(p), \qquad p \in S_2, \tag{1.6.3}$$

where

$$K(p,q) = \left[R_2^2(p) - R_1^2(q)\right] \frac{\partial G(p,q)}{\partial n_q},$$

$$f(p) = \Phi_2(p) - \int_{S_1} \Phi_1(q) \frac{\partial G(p,q)}{\partial n_q}\, \mathrm{d}_q S_1.$$

The Euler equation corresponding to the Tikhonov functional $M^\alpha[\mu_2, f_\delta]$ of equation (1.6.3) for $\mu_2 = z$, $p = x$, $q = s$ is as follows

$$\alpha\Big\{C_0 z - C_1 \frac{\mathrm{d}^2 z}{\mathrm{d}s^2}\Big\} - \int_0^{2\pi} \bar{D}(s,t) z(t)\, \mathrm{d}t = g_\delta(s), \tag{1.6.4}$$

where

$$C_0 = \mathrm{const} \geq 0, \quad C_1 = \mathrm{const} > 0, \quad \bar{D}(s,t) = \int_0^{2\pi} D(x,s) D(x,t)\, \mathrm{d}x,$$

$$D(x,s) = K(x,s)\sqrt{R_1^2(x) + R_1'^2(x)}, \qquad g_\delta(s) = \int_0^{2\pi} D(x,s) f_\delta(x)\, \mathrm{d}x,$$

$$\int_0^{2\pi} [f_\delta(x) - f_T(x)]^2\, \mathrm{d}x \leq \delta^2, \qquad \delta < \int_0^{2\pi} f_\delta^2(x)\, \mathrm{d}x,$$

$$z(0) = z(2\pi), \qquad z'(0) = z'(2\pi). \tag{1.6.5}$$

Analogously to (Tikhonov and Arsenin, 1977), we can show that if α is chosen, z_α will be an approximate solution. Note that in Tikhonov and Arsenin (1977) three approaches to discretization of problems of the type (1.6.4) are presented. We shall show here one more approach. It is easily seen that (1.6.4) is equivalent to the integral equation

$$z_{\alpha\delta}(s) = \frac{1}{\alpha} \int_0^{2\pi} \Phi(s,t) z_{\alpha\delta}(t)\, \mathrm{d}t + F_\delta(s), \tag{1.6.6}$$

where

$$\Phi(s,t) = \int_0^{2\pi} \Gamma(\theta, s) \bar{D}(\theta, t)\, \mathrm{d}\theta, \qquad F_\delta(s) = -\frac{1}{\alpha} \int_0^{2\pi} \Gamma(\theta, s) g_\delta(\theta)\, \mathrm{d}\theta.$$

Here Γ is the periodic Green function of the differential part of (1.6.4):

$$\Gamma(\theta, s) = \frac{1}{2(1 - e^{2\pi a})} \begin{cases} e^{a\theta + as + 2\pi a} + e^{as - a\theta}, & 0 \le \theta \le s, \\ e^{as - a\theta + 2\pi a} + e^{a\theta - as}, & s \le \theta \le 2\pi, \end{cases} \qquad a = \frac{C_0}{C_1}.$$

Evidently, Φ is a continuous periodic kernel. It was shown (see, for example, Tricomi, 1957) that discretization of equation (1.6.6) results in an integral equation with a degenerate kernel $\Phi^\varepsilon(s, t)$

$$Z^\varepsilon_{\alpha\delta}(s) = \frac{1}{\alpha} \int_0^{2\pi} \Phi^\varepsilon(s, t) Z^\varepsilon_{\alpha\delta}(t)\, dt + F_\delta(s), \tag{1.6.7}$$

where

$$\|\Phi(s, t) - \Phi^\varepsilon(s, t)\|_{C(0,2\pi)} < \varepsilon.$$

The following estimate holds

$$\begin{aligned} &\|Z_{\alpha\delta}(s) - Z^\varepsilon_{\alpha\delta}(s)\| \\ &\quad \le \frac{\varepsilon}{\alpha} \cdot \frac{\Sigma(L)\Sigma'(L) + L\left[\Sigma'^2(L) + \Sigma(L)\Sigma''(L)\right]}{|D^*(1/\alpha)|\left[D^*(1/\alpha) - (\varepsilon/\alpha)\Sigma'(L)\right]} \max |F_\delta(s)|, \end{aligned} \tag{1.6.8}$$

where

$$\Sigma(z) = \sum_{m=1}^{\infty} a_m z^m, \qquad a_m = \frac{\sqrt{m}}{m!};$$

$D^*(1/\alpha)$ is the Fredholm function for the kernel Φ^ε and

$$L = \frac{1}{\alpha}[\max |\Phi| + \varepsilon].$$

We may even replace the function $\Sigma(z)$ and its derivatives by the corresponding derivatives of the elementary function

$$\Sigma_0(z) = (1 + z) \exp\Big(\frac{1}{2} e z^2\Big)$$

majorizing (in the Cauchy sense) the function $\Sigma(z)$. Equation (1.6.7) is reduced to a system of linear algebraic equations. Efficient methods of solution of such systems are presented in Tikhonov and Arsenin (1977). In that monograph several equations for determining the regularization parameter α are given. In our case, we find α by the scheme of Lavrent'ev (1973, p. 27). Formulas (1.6.5), (1.6.6), and (1.6.8) yield the estimate

$$\|Z^\varepsilon_{\alpha\delta}(s) - Z_T(s)\|_{C(0,2\pi)} \le \omega(\alpha, \varepsilon, \delta), \tag{1.6.9}$$

where

$$\omega(\alpha, \varepsilon, \delta) = K_0 \cdot \frac{\delta}{\alpha} + \|Z_\alpha(s) - Z_T(s)\| + \frac{\varepsilon}{\alpha} \cdot \frac{\Sigma(L)\Sigma'(L) + L\left[\Sigma'^2(L) + \Sigma(L)\Sigma''(L)\right]}{|D^*(1/\alpha)|\left[D^*(1/\alpha) - (\varepsilon/\alpha)\Sigma'(L)\right]}\|F_\delta\|,$$

and K_0 is a positive constant depending on C_0, C_1, $\max R_1$, $\max |R_1'|$.

For convergence of $Z_{\alpha\delta}^\varepsilon$ to the exact solution $Z_T(s)$ as $\varepsilon, \delta \to 0$ it suffices to take

$$\alpha = \sqrt{\delta}, \qquad \varepsilon = \delta.$$

For fixed accuracy of approximate data, the value of α, when $\inf \omega(\alpha, \varepsilon, \delta)$ is achieved, will be optimal in the sense of the estimate (1.6.9) (see Lavrent'ev, 1973, p. 27).

Chapter 2.

The second internal boundary value problem

We need to find a function $u(p)$ satisfying conditions (1)–(3). We shall seek this function as the sum of a simple layer potential and a logarithmic potential of the domain D_1

$$u(p) = \int_{S_1} \rho_1(q) \ln |p-q| \, \mathrm{d}_q S_1 + \int_{D_1} \omega(Q) \ln |Q-p| \, \mathrm{d}Q, \tag{2.0.1}$$

where

$$\omega(Q) = \int_{S_1} \rho_2(q) \ln |Q-q| \, \mathrm{d}_q S_1.$$

By conditions (2) for $j = 2$ and (3), we obtain from (2.0.1) the following system of integral Fredholm equations of third kind

$$\begin{aligned} -\pi\rho_1(p) + \int_{S_1} \rho_1(q) \frac{\partial}{\partial n_p} \ln |p-q| \, \mathrm{d}_q S_1 & \\ + \int_{S_1} \rho_2(q) \frac{\partial K(p,q)}{\partial n_p} \, \mathrm{d}_q S_1 &= \Phi_1(p), \quad p \in S_1, \\ \int_{S_1} \rho_1(q) \ln |p-q| \, \mathrm{d}_q S_1 + \int_{S_1} \rho_2(q) K(p,q) \, \mathrm{d}_q S_1 &= \Phi_2(p), \quad p \in S_2, \end{aligned} \tag{2.0.2}$$

where

$$K(p,q) = \int_{D_1} \ln |Q-p| \ln |Q-q| \, \mathrm{d}Q.$$

For IBVP 2 we can prove the uniqueness theorems analogous to Theorems 1.0.1, 1.0.2 and consider an example of nonuniqueness similar to Section 1.1.

Theorem 2.0.1. *If S_2 is a circle, then the problem* (1)–(3) *is uniquely completely solvable.*

The proof is just similar to that of Theorem 1.2.1.

2.1. THE SECOND INTERNAL BOUNDARY VALUE PROBLEM IN A CONVEX DOMAIN

Case A. Let

$$D_1 = \{(r,\varphi) \mid 0 \le r \le R_1,\ 0 \le \varphi \le 2\pi\},$$
$$D_2 = \{(r,\varphi) \mid 0 \le r \le R_2,\ 0 \le \varphi \le 2\pi\},$$
$$S_1 = \partial D_1, \qquad S_2 = \partial D_2, \qquad 0 < R_2 < R_1.$$

Problem 2.1.1. We need to find the function $u(r,\varphi)$ in D_1 satisfying the conditions

$$\Delta^2 u(r,\varphi) = 0 \quad \text{in} \quad D_1, \tag{2.1.1}$$

$$\left.\frac{\partial u(r,\varphi)}{\partial r}\right|_{S_1} = \varphi_1(\varphi), \tag{2.1.2}$$

$$u(r,\varphi)\big|_{S_2} = \varphi_2(\varphi), \tag{2.1.3}$$

where Δ is the Laplace operator and φ_j, $j = 1,2$, are given functions.

The following representation holds (Tikhonov and Samarskii, 1990)

$$u(r,\varphi) = u_1(r,\varphi) + (r^2 - R_2^2)u_2(r,\varphi), \tag{2.1.4}$$

where $u_j(r,\varphi)$, $j = 1,2$, are harmonic functions.

By means of representation (2.1.4) we can reduce the problem (2.1.1)–(2.1.3) to the following problems

$$\Delta u_1(r,\varphi) = 0 \quad \text{in} \quad D_1, \tag{2.1.5}$$

$$u_1(r,\varphi)|_{S_2} = \varphi_2(\varphi), \tag{2.1.6}$$

$$\Delta u_2(r,\varphi) = 0 \quad \text{in} \quad D_1, \tag{2.1.7}$$

$$\frac{\partial u_2(r,\varphi)}{\partial r}\Big|_{S_1} + hu_2(r,\varphi)\big|_{S_1} = f(\varphi), \tag{2.1.8}$$

where

$$h = \frac{2R_1}{R_1^2 - R_2^2}, \qquad f(\varphi) = \Big[\varphi_1(\varphi) - \frac{\partial u_1(r,\varphi)}{\partial r}\Big](R_1^2 - R_2^2)^{-1}.$$

The problem (2.1.5), (2.1.6) is the classical ill-posed problem of harmonic (analytic) continuation. This means that there is no continuous dependence of the solution on the data (Lavrent'ev, 1962, 1973). Therefore, the problem (2.1.1)–(2.1.3) is ill-posed.

Uniqueness of solution of the problem (2.1.1)–(2.1.3) follows from representation (2.1.4) and from uniqueness of solutions of the problems (2.1.5), (2.1.6) and (2.1.7), (2.1.8) (Lavrent'ev, 1981).

The following theorem characterizing the stability of solution of the problem (2.1.1)–(2.1.3) holds.

Theorem 2.1.1. *If a function $u(r,\varphi)$ satisfies the relations*

$$\|u(R_1,\varphi)\|_{L_2(-\pi,\pi)} + \|\Delta u(R_1,\varphi)\|_{L_2(-\pi,\pi)} \le M, \tag{2.1.9}$$

$$\Big\|\frac{\partial u(R_1,\varphi)}{\partial r}\Big\|_{L_2(-\pi,\pi)} \le \varepsilon_1, \tag{2.1.10}$$

$$\|u(R_2,\varphi)\|_{L_2(-\pi,\pi)} \le \varepsilon_2, \tag{2.1.11}$$

$$\int_{-\pi}^{\pi} u(R_2,\varphi)\,\mathrm{d}\varphi = 0, \tag{2.1.12}$$

then the inequality

$$\|u(r,\varphi)\|_{L_2(-\pi,\pi)} \le \Big(1 + \frac{|r^2 - R_2^2|}{|R_1^2 - R_2^2|}\Big)\frac{M_1R_1}{\lambda(\varepsilon_2)}\Big(\frac{r}{R_1}\Big)^{\lambda(\varepsilon_2)} + \frac{|r^2 - R_2^2|}{|R_1^2 - R_2^2|}\,\frac{\varepsilon_1}{h} \tag{2.1.13}$$

holds true, where

$$h = \frac{2R_1}{R_1^2 - R_2^2}, \qquad M_1 \le \frac{M}{2(1+h)},$$

and $\lambda(\varepsilon_2)$ is the root of the equation

$$\lambda \ln\frac{R_1}{R_2} + \ln\lambda = \ln\frac{M_1R_1}{\varepsilon_2}. \tag{2.1.14}$$

We shall prove Theorem 2.1.1 later.

Note that, by (2.1.14), $\lambda(\varepsilon_2) \to \infty$ as $\varepsilon_2 \to 0$; more precisely, for ε_2 sufficiently small, we have

$$\lambda > \left[1 + \ln \frac{R_1}{R_2}\right]^{-1} \ln \frac{M_1 R_1}{\varepsilon_2}.$$

We consider now the problem (2.1.5), (2.1.6). The theorem below characterizes the stability of solution of the problem (2.1.5), (2.1.6).

Theorem 2.1.2. *If the function $u_1(r, \varphi)$ satisfies the relations*

$$\left\| \frac{\partial u_1(R_1, \varphi)}{\partial r} \right\|_{L_2(-\pi,\pi)} \le M, \tag{2.1.15}$$

$$\|u_1(R_2, \varphi)\|_{L_2(-\pi,\pi)} \le \varepsilon_2, \tag{2.1.16}$$

$$\int_{-\pi}^{\pi} u_1(R_2, \varphi)\, d\varphi = 0, \tag{2.1.17}$$

then the inequality

$$\|u_1(r, \varphi)\|_{L_2(-\pi,\pi)} \le \frac{M_1 R_1}{\lambda(\varepsilon_2)} \left(\frac{r}{R_1}\right)^{\lambda(\varepsilon_2)} \tag{2.1.18}$$

holds, where $\lambda(\varepsilon)$ is the root of equation (2.1.14).

Proof. The solution to the problem (2.1.5), (2.1.6) can be written in the form

$$u_1(r, \varphi) = \frac{a_0}{2} + \sum_{k=1}^{\infty} \frac{r^k}{R_2^k} (a_k \cos k\varphi + b_k \sin k\varphi). \tag{2.1.19}$$

Introduce the notations

$$\sqrt{a_k^2 + b_k^2} = C_k, \qquad \frac{a_0}{\sqrt{2}} = C_0.$$

Then (2.1.15)–(2.1.17) and (2.1.19) imply

$$\|u_1(r, \varphi)\|^2_{L_2(-\pi,\pi)} = \frac{1}{\pi} \int_{-\pi}^{\pi} u_1^2(r, \varphi)\, d\varphi = \sum_{k=0}^{\infty} \left(\frac{r}{R_2}\right)^{2k} C_k^2, \tag{2.1.20}$$

$$\sum_{k=0}^{\infty} k^2 \left(\frac{r}{R_2}\right)^{2k} C_k^2 \le M_1^2 R_1^2, \tag{2.1.21}$$

$$\sum_{k=0}^{\infty} C_k^2 \leq \varepsilon_2^2. \tag{2.1.22}$$

We now estimate (2.1.20) taking into account (2.1.21) and (2.1.22).

The sum on the right-hand side of (2.1.20) achieves its conditional maximum in the case when $C_k = 0$, $k \neq i, m$, where i, m $(i < m)$ are some numbers; and C_i, C_m satisfy one of three relations

$$\begin{cases} C_i^2 + C_m^2 = \varepsilon_2^2, \\ i^2\left(\dfrac{R_1}{R_2}\right)^{2i} C_i^2 + m^2\left(\dfrac{R_1}{R_2}\right)^{2m} C_m^2 = M_1^2 R_1^2, \end{cases} \tag{2.1.23}$$

$$C_i = 0, \tag{2.1.24}$$

$$C_m = 0. \tag{2.1.25}$$

Suppose that (2.1.23) holds. Then

$$C_i^2 = \left[m^2\left(\frac{R_1}{R_2}\right)^{2m} - i^2\left(\frac{R_1}{R_2}\right)^{2i}\right]^{-1}\left[\varepsilon_2^2 m^2\left(\frac{R_1}{R_2}\right)^{2m} - M_1^2 R_1^2\right], \tag{2.1.26}$$

$$C_m^2 = \left[m^2\left(\frac{R_1}{R_2}\right)^{2m} - i^2\left(\frac{R_1}{R_2}\right)^{2i}\right]^{-1}\left[M_1^2 R_1^2 - \varepsilon_2^2 i^2\left(\frac{R_1}{R_2}\right)^{2i}\right], \tag{2.1.27}$$

which yields

$$\frac{M_1 R_1}{m(R_1/R_2)^m} \leq \varepsilon_2 \leq \frac{M_1 R_1}{i(R_1/R_2)^i}. \tag{2.1.28}$$

In virtue of (2.1.26)–(2.1.28), we obtain

$$\|u_1(r,\varphi)\|^2_{L_2(-\pi,\pi)} \leq \frac{M_1^2 R_1^2}{\lambda^2(\varepsilon_2)}\left(\frac{r}{R_1}\right)^{2\lambda(\varepsilon_2)}, \tag{2.1.29}$$

where $\lambda(\varepsilon_2)$ is the root of equation (2.1.14).

Suppose now that relation (2.1.24) holds. Then,

$$\|u_1(r,\varphi)\|^2_{L_2(-\pi,\pi)} \leq \left(\frac{r}{R_1}\right)^{2m} C_m^2.$$

Relations (2.1.21) and (2.1.22) yield

$$m^2\left(\frac{R_1}{R_2}\right)^{2m} C_m^2 \leq M_1^2 R_1^2, \qquad C_m^2 \leq \varepsilon_2^2.$$

Hence it follows that

$$\|u_1(r,\varphi)\|^2_{L_2(-\pi,\pi)} \le \frac{M_1^2 R_1^2}{\lambda^2(\varepsilon_2)}\Big(\frac{r}{R_1}\Big)^{2\lambda(\varepsilon_2)} \tag{2.1.30}$$

where $\lambda(\varepsilon_2)$ is the root of equation (2.1.14).

If relation (2.1.25) holds, inequality (2.1.30) is obtained analogously.

The statement of the theorem follows from (2.1.29) and (2.1.30). □

Consider now the family of linear operators B_n dependent on the integer parameter and defined as follows

$$B_n\varphi_2(\varphi) = \frac{a_0}{2} + \sum_{k=1}^{n}\Big(\frac{r}{R_2}\Big)^k (a_k\cos k\varphi + b_k \sin k\varphi), \tag{2.1.31}$$

where

$$a_0 = \frac{1}{\pi}\int_{-\pi}^{\pi}\varphi_2(\varphi)\,\mathrm{d}\varphi, \qquad a_k = \frac{1}{\pi}\int_{-\pi}^{\pi}\varphi_2(\varphi)\cos k\varphi\,\mathrm{d}\varphi,$$
$$b_k = \frac{1}{\pi}\int_{-\pi}^{\pi}\varphi_2(\varphi)\sin k\varphi\,\mathrm{d}\varphi.$$

The family of operators B_n will be regularizing (Lavrent'ev, 1973, p. 25) relative to the problem (2.1.5), (2.1.6) if we consider $\varphi_2(\varphi)$ and the solution $u_1(r,\varphi)$ as elements of the Hilbert spaces L_2. Really, the operators of determining the Fourier coefficients of $\varphi_2(\varphi)$ and of finite summation are continuous. Therefore, the operators B_n are continuous. Convergence of the sequence $B_n\varphi_2(\varphi)$ to the solution $u_1(r,\varphi)$ follows from the representation of solution (2.1.19) and from convergence of the Fourier series in the space L_2.

Now let us estimate the efficiency of application of this family to construction of an approximate solution by approximate data. We assume that the problem (2.1.5)–(2.1.8) is set conditionally well-posed, and the well-posedness set is defined by inequality (2.1.15).

Suppose that we know $\varphi_2(\varphi)$ to within ε, i. e., we know $\varphi_{2\varepsilon}(\varphi)$ such that

$$\|\varphi_2(\varphi) - \varphi_{2\varepsilon}(\varphi)\|_{L_2(-\pi,\pi)} \le \varepsilon.$$

As an approximate solution of (2.1.5), (2.1.6) we take the function

$$u_{1n\varepsilon}(r,\varphi) = B_n\varphi_{2\varepsilon}(\varphi) = \frac{\bar a_0}{2} + \sum_{k=1}^{n}\Big(\frac{r}{R_2}\Big)^k(\bar a_k\cos k\varphi + \bar b_k\sin k\varphi), \tag{2.1.32}$$

where

$$\bar{a}_0 = \frac{1}{\pi}\int_{-\pi}^{\pi} \varphi_{2\varepsilon}(\varphi)\,d\varphi, \qquad \bar{a}_k = \frac{1}{\pi}\int_{-\pi}^{\pi} \varphi_{2\varepsilon}(\varphi)\cos k\varphi\,d\varphi,$$

$$\bar{b}_k = \frac{1}{\pi}\int_{-\pi}^{\pi} \varphi_{2\varepsilon}(\varphi)\sin k\varphi\,d\varphi.$$

Estimate the difference between $u_{1n\varepsilon}(r,\varphi)$ and $u_1(r,\varphi)$

$$\begin{aligned} &\|u_1(r,\varphi) - u_{1n\varepsilon}(r,\varphi)\|_{L_2(-\pi,\pi)} \\ &\quad = \|B_n\varphi_{2\varepsilon}(\varphi) - u_1(r,\varphi)\|_{L_2(-\pi,\pi)} \\ &\quad \le \|B_n\varphi_{2\varepsilon}(\varphi) - B_n\varphi_2(\varphi)\|_{L_2(-\pi,\pi)} + \|B_n\varphi_2(\varphi) - u_1(r,\varphi)\|_{L_2(-\pi,\pi)} \\ &\quad \le \|B_n\|_{L_2(-\pi,\pi)}\varepsilon + \|B_n\varphi_2(\varphi) - u_1(r,\varphi)\|_{L_2(-\pi,\pi)}. \end{aligned} \tag{2.1.33}$$

Relations (2.1.19) and (2.1.31) yield

$$\|B_n\|_{L_2(-\pi,\pi)} = \left(\frac{r}{R_1}\right)^n, \tag{2.1.34}$$

$$\|B_n\varphi_2(\varphi) - u_1(r,\varphi)\|_{L_2(-\pi,\pi)} = \sum_{k=n+1}^{\infty}\left(\frac{r}{R_2}\right)^{2k} C_k^2. \tag{2.1.35}$$

The sum on the right-hand side of (2.1.35) under the condition (2.1.21) achieves its maximal value in the case when

$$C_k = 0, \qquad k \ne n+1, \qquad C_{n+1} = \frac{M_1R_1}{n+1}\left(\frac{R_2}{R_1}\right)^{n+1},$$

$$\|B_n\varphi_2(\varphi) - u_1(r,\varphi)\|_{L_2(-\pi,\pi)} \le \frac{M_1R_1}{n+1}\left(\frac{r}{R_1}\right)^{n+1}. \tag{2.1.36}$$

Therefore,

$$\|u_1(r,\varphi) - u_{1n\varepsilon}(r,\varphi)\|_{L_2(-\pi,\pi)} \le \left(\frac{r}{R_2}\right)^n \varepsilon + \frac{M_1R_1}{n+1}\left(\frac{r}{R_1}\right)^{n+1}. \tag{2.1.37}$$

We consider now the problem (2.1.7), (2.1.8). Let

$$\varphi_1(\varphi) = \frac{\alpha_0}{2} + \sum_{k=1}^{\infty}(\alpha_k\cos k\varphi + \beta_k\sin k\varphi).$$

Then the solution of the problem (2.1.7), (2.1.8) can be written as follows

$$u_2(r,\varphi) = \bar{u}_2(r,\varphi) - \tilde{u}_2(r,\varphi), \tag{2.1.38}$$

where

$$\bar{u}_2(r,\varphi) = \frac{1}{R_1^2 - R_2^2}\left\{\frac{\alpha_0}{2h} + \sum_{k=1}^{\infty} \frac{r^k}{(k+hR_1)R_1^{k-1}}(\alpha_k \cos k\varphi + \beta_k \sin k\varphi)\right\}, \tag{2.1.39}$$

$$\tilde{u}_2(r,\varphi) = \frac{1}{R_1^2 - R_2^2}\sum_{k=1}^{\infty} \frac{r^k(\alpha_k \cos k\varphi + \beta_k \sin k\varphi)}{(1+(hR_1)/k)R_2^k}. \tag{2.1.40}$$

Theorem 2.1.2 and (2.1.40) yield

$$\|\tilde{u}_2(r,\varphi)\|_{L_2(-\pi,\pi)} \le \frac{1}{R_1^2 - R_2^2}\frac{M_1 R_1}{\lambda(\varepsilon_2)}\left(\frac{r}{R_1}\right)^{\lambda(\varepsilon_2)}, \tag{2.1.41}$$

where $\lambda(\varepsilon_2)$ is the root of equation (2.1.14).

Proof of Theorem 2.1.1. Relations (2.1.17), (2.1.19), and (2.1.40) yield

$$\|u_1(R_1,\varphi)\|_{L_2(-\pi,\pi)} \le M_1 R_1 \left\|\frac{\partial u_1(R_1,\varphi)}{\partial r}\right\|_{L_2(-\pi,\pi)} \le M, \tag{2.1.42}$$

$$\|\tilde{u}_2(r,\varphi)\|_{L_2(-\pi,\pi)} \le \frac{M_1 R_1}{R_1^2 - R_2^2}, \qquad \left\|\frac{\partial \tilde{u}_2(R_1,\varphi)}{\partial r}\right\|_{L_2(-\pi,\pi)} \le \frac{M_1}{R_1^2 - R_2^2}.$$

Representation (2.1.4) implies

$$\Delta u(r,\varphi) = 4u_2(r,\varphi) + 4r\frac{\partial u_2(r,\varphi)}{\partial r}. \tag{2.1.43}$$

Relations (2.1.4), (2.1.38), (2.1.42), and (2.1.43) yield

$$\|u(R_1,\varphi)\|_{L_2(-\pi,\pi)} + \|\Delta u(R_1,\varphi)\|_{L_2(-\pi,\pi)}$$
$$\le 2R_1\left(1 + \frac{4}{R_1^2 - R_2^2}\right)M_1 + \frac{1}{h}\left[1 + \frac{4(1+R_1 h)}{R_1^2 - R_2^2}\right]\|\varphi_1(\varphi)\|_{L_2(-\pi,\pi)}.$$

We set

$$M = 2R_1\left(1 + \frac{4}{R_1^2 - R_2^2}\right)M_1 + \frac{1}{h}\left[1 + \frac{4(1+R_1 h)}{R_1^2 - R_2^2}\right]\|\varphi_1(\varphi)\|_{L_2(-\pi,\pi)},$$

which implies

$$M_1 \le \frac{M}{2(1+h)}.$$

Now, the statement of Theorem 2.1.1 follows from Theorem 2.1.2 and relations (2.1.4), (2.1.10), (2.1.38), and (2.1.41). □

Suppose that we know $\varphi_1(\varphi)$ and $\varphi_2(\varphi)$ with accuracy δ and ε, respectively. This means that

$$\|\varphi_{1\delta}(\varphi) - \varphi_1(\varphi)\|_{L_2(-\pi,\pi)} \leq \delta, \quad \|\varphi_{2\varepsilon}(\varphi) - \varphi_2(\varphi)\|_{L_2(-\pi,\pi)} \leq \varepsilon. \tag{2.1.44}$$

We take as an approximate solution to the problem (2.1.1)–(2.1.3) the function

$$u_{n\varepsilon\delta}(r,\varphi) = \bar{u}_{1n\varepsilon}(r,\varphi) + (r^2 - R_2^2)u_{2n\varepsilon\delta}(r,\varphi), \tag{2.1.45}$$

where $\bar{u}_{1n\varepsilon}(r,\varphi)$ is the approximate solution of the problem (2.1.5), (2.1.6); and

$$u_{2n\varepsilon\delta}(r,\varphi) = \bar{u}_{2n\delta}(r,\varphi) - \tilde{u}_{2n\varepsilon}(r,\varphi), \tag{2.1.46}$$

$$\bar{u}_{2n\delta}(r,\varphi) = \frac{1}{R_1^2 - R_2^2}\left\{\frac{\bar{\alpha}_0}{2h} + \sum_{k=1}^{n} \frac{r^k}{(k + hR_1)R_1^{k-1}}(\bar{\alpha}_k \cos k\varphi + \bar{\beta}_k \sin k\varphi)\right\}, \tag{2.1.47}$$

$$\tilde{u}_{2n\varepsilon}(r,\varphi) = \frac{1}{R_1^2 - R_2^2}\left\{\sum_{k=1}^{n} \frac{r^k}{(k + hR_1)R_2^{k-1}}(\bar{\alpha}_k \cos k\varphi + \bar{\beta}_k \sin k\varphi)\right\}. \tag{2.1.48}$$

Here

$$\bar{\alpha}_0 = \frac{1}{\pi}\int_{-\pi}^{\pi} \varphi_{1\delta}(\varphi)\,\mathrm{d}\varphi, \qquad \bar{\alpha}_k = \frac{1}{\pi}\int_{-\pi}^{\pi} \varphi_{1\delta}(\varphi)\cos k\varphi\,\mathrm{d}\varphi,$$

$$\bar{\beta}_k = \frac{1}{\pi}\int_{-\pi}^{\pi} \varphi_{1\delta}(\varphi)\sin k\varphi\,\mathrm{d}\varphi.$$

From (2.1.40) and (2.1.48), taking into account (2.1.37), we find

$$\|\tilde{u}_{2n\varepsilon}(r,\varphi) - \tilde{u}_2(r,\varphi)\|_{L_2(-\pi,\pi)} \leq \frac{1}{R_1^2 - R_2^2}\left[\left(\frac{r}{R_2}\right)^n \varepsilon + \frac{M_1 R_1}{n+1}\left(\frac{r}{R_1}\right)^{n+1}\right]. \tag{2.1.49}$$

Now, relations (2.1.4), (2.1.9), (2.1.37), (2.1.38) and (2.1.42)–(2.1.49) yield

$$\|u(r,\varphi) - u_{n\varepsilon\delta}(r,\varphi)\|_{L_2(-\pi,\pi)} \leq \omega(n,\varepsilon,\delta), \tag{2.1.50}$$

$$\omega(n,\varepsilon,\delta) = \left(1 + \frac{|r^2 - R_2^2|}{R_1^2 - R_2^2}\right)\left[\left(\frac{r}{R_2}\right)^n \varepsilon + \frac{M_1 R_1}{n+1}\left(\frac{r}{R_1}\right)^{n+1}\right]$$

$$+\frac{|r^2-R_2^2|}{4R_1}\Big[2\delta+R_1(M+2hM_1)\Big(\frac{r}{R_1}\Big)^{n+1}\Big].$$

Note that the efficiency of regularization depends on the choice of the regularization parameter n, which can be obtained from the equation

$$\Big(\frac{r}{R_2}\Big)^n\varepsilon=\frac{M_1R_1}{n+1}\Big(\frac{r}{R_1}\Big)^{n+1},$$

which means that, for ε sufficiently small,

$$n>\Big[1+\ln\frac{r}{R_1}\Big]^{-1}\ln\frac{M_1r}{\varepsilon},$$

or from the equation

$$2\delta=R_1(M+2hM_1)\Big(\frac{r}{R_1}\Big)^{n+1},$$

i. e.,

$$n+1=\Big[\ln\frac{r}{R_1}\Big]^{-1}\ln\frac{2\delta}{R_1(M+2hM_1)}.$$

For fixed accuracy ε, δ of approximate data, the value of the parameter n, for which $\inf\omega(n,\varepsilon,\delta)$ is achieved, will be optimal in the sense of the estimate (2.1.50).

Case B. We suppose here that D_1 and D_2 ($D_1\supset\overline{D_2}$) are domains on the plane bounded by the curves S_1 and S_2, respectively. We suppose also that $S_1\colon\ r=\rho(\varphi)$ is a convex closed curve with continuous curvature, and $S_2\colon\ r=R_2=\text{const}$.

Problem 2.1.2. We need to find the function $u(r,\varphi)$ in D_1 satisfying the conditions

$$\Delta^2u(r,\varphi)=0\qquad\text{in}\quad D_1,\tag{2.1.51}$$

$$\frac{\partial u(r,\varphi)}{\partial\nu}\Big|_{S_1}=\varphi_1(\varphi),\tag{2.1.52}$$

$$u(r,\varphi)|_{S_2}=\varphi_2(\varphi),\tag{2.1.53}$$

where Δ is the Laplace operator; ν is the external normal to S_1; and $\varphi_j(\varphi)$, $j=1,2$, are given functions.

Using representation (2.1.4), we reduce the problem (2.1.51)–(2.1.53) to the problems

$$\Delta u_1(r,\varphi) = 0 \quad \text{in} \quad D_1, \tag{2.1.54}$$

$$u_1(r,\varphi)|_{S_2} = \varphi_2(\varphi), \tag{2.1.55}$$

$$\Delta u_2(r,\varphi) = 0 \quad \text{in} \quad D_1, \tag{2.1.56}$$

$$\left[\frac{\partial u_2(r,\varphi)}{\partial \nu} + \frac{2r\cos(r,\nu)}{r^2 - R_2^2} u_2(r,\varphi)\right]\Big|_{S_1} = \varphi_3(\varphi), \tag{2.1.57}$$

where

$$\varphi_3(\varphi) = \left[\varphi_1(\varphi) - \frac{\partial u_1(r,\varphi)}{\partial \nu}\Big|_{S_1}\right](\rho^2(\varphi) - R_2^2)^{-1}. \tag{2.1.58}$$

The problem (2.1.54), (2.1.55) is a classically ill-posed problem of harmonic continuation (see Lavrent'ev, 1962, 1973). Therefore, the problem (2.1.51)–(2.1.53) is ill-posed.

Solution of the problem (2.1.56), (2.1.57) is unique (Koshlyakov *et al.*, 1970), since

$$\frac{2r\cos(r,\nu)}{r^2 - R_2^2}\Big|_{S_1} > 0.$$

Since the problems (2.1.54), (2.1.55) and (2.1.56), (2.1.57) are solvable uniquely, therefore, representation (2.1.4) implies the uniqueness of solution to the problem (2.1.51)–(2.1.53).

The next theorem characterizes the stability of solution to the problem (2.1.51)–(2.1.53).

Theorem 2.1.3. *If the function $u(r,\varphi)$ satisfies the relations*

$$\|u(r,\varphi)|_{S_1}\|_{L_2(-\pi,\pi)} + \left\|\frac{\partial u(r,\varphi)}{\partial \nu}\Big|_{S_1}\right\|_{L_2(-\pi,\pi)} \le M, \tag{2.1.59}$$

$$\left\|\frac{\partial u(r,\varphi)}{\partial \nu}\Big|_{S_1}\right\|_{L_2(-\pi,\pi)} \le \varepsilon_1, \tag{2.1.60}$$

$$\|u(r,\varphi)|_{S_2}\|_{L_2(-\pi,\pi)} \le \varepsilon_2, \tag{2.1.61}$$

$$\int_{-\pi}^{\pi} u(R_2,\varphi)\,\mathrm{d}\varphi = 0, \tag{2.1.62}$$

then the inequality

$$\|u(r,\varphi)\|_{L_2(-\pi,\pi)} \le \frac{M_1 R_1}{\lambda(\varepsilon_2)}\left(\frac{r}{R_1}\right)^{\lambda(\varepsilon_2)}$$

$$+ |r^2 - R_2^2| a\bar{h}(r) \Big[\varepsilon_1 + \frac{M_1 R_1}{R} \Big(\frac{R}{R_1}\Big)^{\lambda(\varepsilon_2)} \Big] \quad (2.1.63)$$

holds true, where

$$R_1 > R = \max_{0<\varphi<2\pi} \rho(\varphi), \qquad M_1 < \frac{M}{R_1 + abL},$$

$$a = \max_{0<\varphi<2\pi} [\rho^2(\varphi) - R_2^2]^{-1}, \qquad b = R^2 - R_2^2,$$

$L = \mathrm{const}$, *$\bar{h}(r)$ is a bounded function, and $\lambda(\varepsilon_2)$ is the root of the equation*

$$\lambda \ln \frac{R_1}{R_2} + \ln \lambda = \ln \frac{M_1 R_1}{\varepsilon_2}.$$

We shall prove this theorem later.

Consider the problem (2.1.5), (2.1.6). Note that this problem is the problem of harmonic continuation into the domain D_{R_1} and $D_{R_1} \supset \overline{D_1}$. Therefore, its solution $u_1(r, \varphi)$ is also a solution of the problem (2.1.54), (2.1.55).

In Subsection "Case A", the conditional well-posedness of the problem (2.1.5), (2.1.6) was studied and an approximate solution was constructed.

The following statement holds.

Theorem 2.1.4. *If the function $u_1(r, \varphi)$ satisfies relations* (2.1.15)–(2.1.17), *then the inequality*

$$\Big\| \frac{\partial u_1(r, \varphi)}{\partial r} \Big\|_{L_2(-\pi,\pi)} < \frac{M_1 R_1}{r} \Big(\frac{r}{R_1}\Big)^{\lambda(\varepsilon_2)} \quad (2.1.64)$$

holds, where $\lambda(\varepsilon_2)$ is the root of equation (2.1.14).

The proof of this theorem is similar to that of Theorem 2.1.2.

From the equality

$$\frac{\partial u_1(r, \varphi)}{\partial \nu} = \frac{\partial u_1(r, \varphi)}{\partial r} \cos(r, \nu),$$

taking into account $\cos^2(r, \nu) \le 1$, we obtain

$$\Big\| \frac{\partial u_1(r, \varphi)}{\partial \nu} \Big\|_{L_2(-\pi,\pi)} \le \Big\| \frac{\partial u_1(r, \varphi)}{\partial r} \Big\|_{L_2(-\pi,\pi)}. \quad (2.1.65)$$

Hence, taking into account (2.1.64), we obtain

$$\left\| \frac{\partial u_1(r,\varphi)}{\partial \nu}\Big|_{S_1} \right\|_{L_2(-\pi,\pi)} \le \left\| \frac{\partial u_1(r,\varphi)}{\partial r}\Big|_{S_1} \right\|_{L_2(-\pi,\pi)} \le \frac{M_1 R_1}{R}\left(\frac{R}{R_1}\right)^{\lambda(\varepsilon_2)}. \tag{2.1.66}$$

Analogously to (2.1.37) we find

$$\left\| \frac{\partial}{\partial r}\left[u_{1n\varepsilon}(r,\varphi) - u_1(r,\varphi)\right] \right\|_{L_2(-\pi,\pi)} \le \frac{n}{r}\left(\frac{r}{R_2}\right)^n \varepsilon + M_1\left(\frac{r}{R_1}\right)^{n+1},$$

where $u_{1n\varepsilon}(r,\varphi)$ is the approximate solution of the problem (2.1.5), (2.1.6). Hence, taking into account (2.1.65), we obtain

$$\left\| \frac{\partial}{\partial \nu}\left[u_{1n\varepsilon}(r,\varphi) - u_1(r,\varphi)\right]\Big|_{S_1} \right\|_{L_2(-\pi,\pi)} \le \frac{n}{R}\left(\frac{R}{R_2}\right)^n \varepsilon + M_1\left(\frac{R}{R_1}\right)^{n+1}. \tag{2.1.67}$$

Let us consider the problem (2.1.56), (2.1.57). We shall seek the solution of this problem as the following potential of simple layer

$$u_2(r,\varphi) = \int_{-\pi}^{\pi} \mu(\psi)\sqrt{\rho^2(\psi) + [\rho'(\psi)]^2} \times \ln \frac{1}{\sqrt{r^2 + \rho^2(\psi) - 2r\rho(\psi)\cos(\varphi - \psi)}}\, d\psi. \tag{2.1.68}$$

Using the theorem on limit values of the normal derivative of the potential (2.1.68), we obtain for $\mu(\varphi)$ the integral equation with a weak singularity

$$\mu(\varphi) + \frac{1}{\pi}\int_{-\pi}^{\pi}\left[\frac{\partial}{\partial \nu}\ln\frac{1}{\sqrt{\rho^2(\varphi) + \rho^2(\psi) - 2\rho(\varphi)\rho(\psi)\cos(\varphi-\psi)}} + \alpha(\varphi)\ln\frac{1}{\sqrt{\rho^2(\varphi) + \rho^2(\psi) - 2\rho(\varphi)\rho(\psi)\cos(\varphi-\psi)}}\right] \times \mu(\psi)\sqrt{\rho^2(\psi) + [\rho'(\varphi)]^2}\, d\psi = \frac{1}{\pi}\varphi_3(\varphi), \tag{2.1.69}$$

where

$$\alpha(\varphi) = \frac{2\rho(\varphi)\cos(r,\nu)}{\rho^2(\varphi) - R_2^2}.$$

Since the problem (2.1.56), (2.1.57) is equivalent to equation (2.1.69), the homogeneous equation corresponding to equation (2.1.69) has only the trivial solution. Therefore, the integral equation (2.1.69) has the unique solution and $\mu(\varphi)$ can be written as follows

$$\mu(\varphi) = \frac{1}{\pi}\varphi_3(\varphi) + \frac{1}{\pi}\int_{-\pi}^{\pi} \Gamma(\varphi,\psi,1/\pi)\varphi_3(\psi)\,d\psi,$$

where Γ is the resolvent of equation (2.1.69).

Let us estimate the norm of $\mu(\varphi)$:

$$\|\mu(\varphi)\|_{L_2(-\pi,\pi)} \le \frac{1}{\pi}\|\varphi_3(\varphi)\|_{L_2(-\pi,\pi)} + \frac{1}{\pi}\Big\|\int_{-\pi}^{\pi} \Gamma(\varphi,\psi,1/\pi)\varphi_3(\psi)\,d\psi\Big\|_{L_2(-\pi,\pi)}.$$

Using the Cauchy - Schwartz - Bunyakovskii inequality, we obtain

$$\Big\|\int_{-\pi}^{\pi} \Gamma(\varphi,\psi,1/\pi)\varphi_3(\psi)\,d\psi\Big\|_{L_2(-\pi,\pi)} \le \Big[\int_{-\pi}^{\pi}\int_{-\pi}^{\pi} \Gamma^2(\varphi,\psi,1/\pi)\,d\varphi\,d\psi\Big]^{1/2}\|\varphi_3(\psi)\|_{L_2(-\pi,\pi)}.$$

Therefore, we have

$$\|\mu(\varphi)\|_{L_2(-\pi,\pi)} \le \frac{1}{\pi}(1+A)\|\varphi_3(\varphi)\|_{L_2(-\pi,\pi)}, \tag{2.1.70}$$

where

$$A = \Big[\int_{-\pi}^{\pi}\int_{-\pi}^{\pi} \Gamma^2(\varphi,\psi,1/\pi)\,d\varphi\,d\psi\Big]^{1/2}.$$

Now we estimate the norm of $u_2(r,\varphi)$. The Cauchy - Schwartz - Bunyakovskii inequality and (2.1.70) yield

$$\|u_2(r,\varphi)\|_{L_2(-\pi,\pi)} \le \bar{h}(r)\|\varphi_3(\varphi)\|_{L_2(-\pi,\pi)}. \tag{2.1.71}$$

Here

$$\bar{h}(r) = \frac{1}{\pi}(1+A)\Big[\int_{-\pi}^{\pi}\int_{-\pi}^{\pi}\Big(\sqrt{\rho^2(\psi)+[\rho'(\psi)]^2} \times \ln\frac{1}{\sqrt{r^2+\rho^2(\psi)-2r\rho(\psi)\cos(\varphi-\psi)}}\Big)^2 d\varphi\,d\psi\Big]^{1/2}.$$

Analogously, we obtain

$$\|u_2(r,\varphi)|_{S_2}\|_{L_2(-\pi,\pi)} \le L\|\varphi_3(\varphi)\|_{L_2(-\pi,\pi)}, \tag{2.1.72}$$

where

$$L = \frac{1}{\pi}(1+A)\Bigg[\int_{-\pi}^{\pi}\int_{-\pi}^{\pi}\Bigg(\sqrt{\rho^2(\psi)+[\rho'(\psi)]^2} \times \ln\frac{1}{\sqrt{\rho^2(\varphi)+\rho^2(\psi)-2\rho(\varphi)\rho(\psi)\cos(\varphi-\psi)}}\Bigg)^2 d\varphi\, d\psi\Bigg]^{1/2}.$$

The norm of $\varphi_3(\varphi)$ is estimated as follows

$$\|\varphi_3(\varphi)\|_{L_2(-\pi,\pi)} \le a\Big[\|\varphi_1(\varphi)\|_{L_2(-\pi,\pi)} + \Big\|\frac{\partial u_1(r,\varphi)}{\partial\nu}\Big|_{S_1}\Big\|_{L_2(-\pi,\pi)}\Big]. \quad (2.1.73)$$

Proof of Theorem 2.1.3. Representation (2.1.4) and formulas (2.1.15), (2.1.65), (2.1.72) and (2.1.73) yield

$$\|u(r,\varphi)|_{S_1}\|_{L_2(-\pi,\pi)} + \Big\|\frac{\partial u(r,\varphi)}{\partial\nu}\Big|_{S_1}\Big\|_{L_2(-\pi,\pi)} \le M_1R_1 + abL\left[\|\varphi_1(\varphi)\|_{L_2(-\pi,\pi)} + M_1\right] + \|\varphi_1(\varphi)\|_{L_2(-\pi,\pi)}.$$

Let

$$M = (R_1 + abL)M_1 + (1+abL)\|\varphi_1(\varphi)\|_{L_2(-\pi,\pi)}.$$

Hence, we obtain

$$M_1 \le \frac{M}{R_1+abL}.$$

Using Theorem 2.1.2 and taking into account (2.1.4), (2.1.59)–(2.1.62), (2.1.66), (2.1.71) and (2.1.73), we prove the statement of Theorem 2.1.3. □

Suppose now that $\varphi_1(\varphi)$ and $\varphi_2(\varphi)$ are known with accuracy δ and ε, respectively; i. e., we know $\varphi_{1\delta}(\varphi)$ and $\varphi_{2\varepsilon}(\varphi)$ such that

$$\|\varphi_{1\delta}(\varphi)-\varphi_1(\varphi)\|_{L_2(-\pi,\pi)} \le \delta, \qquad \|\varphi_{2\varepsilon}(\varphi)-\varphi_2(\varphi)\|_{L_2(-\pi,\pi)} \le \varepsilon. \quad (2.1.74)$$

As an approximate solution of the problem (2.1.51)–(2.1.53) we can take the function

$$u_{n\delta\varepsilon}(r,\varphi) = u_{1n\varepsilon}(r,\varphi) + (r^2-R_2^2)u_{2n\delta\varepsilon}(r,\varphi),$$

where $u_{1n\varepsilon}$ is the approximate solution of the problem (2.1.56), (2.1.57).

Now we estimate the difference between $u_{n\delta\varepsilon}(r,\varphi)$ and $u(r,\varphi)$

$$\begin{aligned}\|u_{n\delta\varepsilon}(r,\varphi) - u(r,\varphi)\|_{L_2(-\pi,\pi)} \\ \le \|u_{1n\varepsilon}(r,\varphi) - u_1(r,\varphi)\|_{L_2(-\pi,\pi)} \\ + |r^2 - R_2^2| \cdot \|u_{2n\delta\varepsilon}(r,\varphi) - u_2(r,\varphi)\|_{L_2(-\pi,\pi)}.\end{aligned} \tag{2.1.75}$$

Formulas (2.1.71) and (2.1.73) yield

$$\begin{aligned}\|u_{2n\delta\varepsilon}(r,\varphi) - u_2(r,\varphi)\|_{L_2(-\pi,\pi)} \\ \le a\bar{h}(r)\Big[\|\varphi_{1\delta}(\varphi) - \varphi_1(\varphi)\|_{L_2(-\pi,\pi)} \\ + \Big\|\frac{\partial}{\partial\nu}[u_{1n\varepsilon}(r,\varphi) - u_1(r,\varphi)]\Big|_{S_1}\Big\|_{L_2(-\pi,\pi)}\Big].\end{aligned} \tag{2.1.76}$$

Now, formulas (2.1.37), (2.1.67), (2.1.74), and (2.1.76) yield

$$\|u_{n\delta\varepsilon}(r,\varphi) - u(r,\varphi)\|_{L_2(-\pi,\pi)} \le \omega(n,\delta,\varepsilon), \tag{2.1.77}$$

where

$$\begin{aligned}\omega(n,\delta,\varepsilon) = \Big[\Big(\frac{r}{R_2}\Big)^n + |r^2 - R_2^2|a\bar{h}(r)\frac{r}{R}\Big(\frac{r}{R_2}\Big)^n\Big]\varepsilon \\ + |r^2 - R_2^2|a\bar{h}(r)\delta + \Big[\frac{R_1}{n+1}\Big(\frac{r}{R_1}\Big)^{n+1} + \Big(\frac{R}{R_1}\Big)^{n+1}\Big]M_1.\end{aligned}$$

Minimizing the right-hand side of (2.1.77) with respect to n, we find the optimal value of the parameter n in the sense of the estimate (2.1.77)

$$n(\delta,\varepsilon) = \inf_n \omega(n,\delta,\varepsilon).$$

2.2. THE SECOND INTERNAL BOUNDARY VALUE PROBLEM IN THE HALF-PLANE

Let

$$\begin{aligned}D_1 = \{(x,y) \mid -\infty < x < \infty,\ 0 < y < \infty\}, \\ D_2 = \{(x,y) \mid -\infty < x < \infty,\ b < y < \infty\}, \qquad b > 0.\end{aligned}$$

Problem 2.2.1. We need to find the function $u(x, y)$ in D_1 satisfying the conditions

$$\Delta^2 u(x, y) = 0 \quad \text{in} \quad D_1, \tag{2.2.1}$$

$$\frac{\partial u(x, 0)}{\partial y} = \varphi_1(x), \tag{2.2.2}$$

$$u(x, b) = \varphi_2(x), \tag{2.2.3}$$

$$u(x, y) = O(1) \quad \text{for} \quad \sqrt{x^2 + y^2} \to \infty. \tag{2.2.4}$$

Here, as before, Δ is the Laplace operator; $\varphi_1(x)$ and $\varphi_2(x)$ are given functions from the class $L_1(-\infty, \infty) \cap L_2(-\infty, \infty)$.

The following representation holds (Tikhonov and Samarskii, 1990):

$$u(x, y) = u_1(x, y) + (y - b)u_2(x, y), \tag{2.2.5}$$

where $u_1(x, y)$ and $u_2(x, y)$ are harmonic functions.

From representation (2.2.5) and the problem conditions it follows that

$$\Delta u_1(x, y) = 0 \quad \text{in} \quad D_1, \tag{2.2.6}$$

$$u_1(x, b) = \varphi_2(x), \tag{2.2.7}$$

$$u_1(x, y) = O(1) \quad \text{for} \quad \sqrt{x^2 + y^2} \to \infty, \tag{2.2.8}$$

$$\Delta u_2(x, y) = 0 \quad \text{in} \quad D_1, \tag{2.2.9}$$

$$\frac{\partial u_2(x, 0)}{\partial y} - \frac{1}{b} u_2(x, 0) = \varphi_3(x), \tag{2.2.10}$$

$$y u_2(x, y) = O(1) \quad \text{for} \quad \sqrt{x^2 + y^2} \to \infty, \tag{2.2.11}$$

where

$$\varphi_3(x) = \frac{1}{b}\Big[\frac{\partial u_1(x, 0)}{\partial y} - \varphi_1(x)\Big].$$

The problem (2.2.6)–(2.2.8) is the ill-posed problem of harmonic continuation from the domain $y > b$ into the strip $0 < y < b$ (Lavrent'ev, 1962, 1973). Therefore, the problem (2.2.1)–(2.2.4) is ill-posed. Note that uniqueness of solution of the problem (2.2.1)–(2.2.4) follows from representation (2.2.5) and from uniqueness of solutions to the problems (2.2.6)–(2.2.8) and (2.2.9)–(2.2.11) (see Koshlyakov *et al.*, 1970).

We shall seek the solution of the problem (2.2.1)–(2.2.4) in the form

$$u(x, y) = \frac{2}{\pi}\int_{-\infty}^{\infty} \frac{y^3 f(s)\,ds}{[(x - s)^2 + y^2]^2} + \frac{1}{\pi}\int_{-\infty}^{\infty} \frac{y^2 g(s)\,ds}{(x - s)^2 + y^2}, \tag{2.2.12}$$

where $f(s)$, $g(s)$ are unknown functions.

The function $u(x, y)$ from (2.2.12) satisfies equation (2.2.1). We need to find the functions $f(s)$ and $g(s)$.

From condition (2.2.2) we find $g(s) = \varphi_1(s)$. As follows from (2.2.3), for determination of the function $f(s)$, we have the integral equation of first kind of convolution type

$$A[x, f(s)] = \int_{-\infty}^{\infty} K(x-s) f(s)\, \mathrm{d}s = \psi(x), \qquad -\infty \le x \le \infty, \quad (2.2.13)$$

where

$$K(x-s) = \frac{2b^3}{\pi}[(x-s)^2 + b^2]^{-2},$$

$$\psi(x) = \varphi_2(x) - \frac{b^2}{\pi}\int_{-\infty}^{\infty} \frac{\varphi_1(s)\, \mathrm{d}s}{(x-s)^2 + b^2}. \quad (2.2.14)$$

Integral equations of first kind of convolution type were considered by Tikhonov and Arsenin (1977).

We see that

$$K(y) \in L_1(-\infty, \infty), \qquad \psi(x) \in L_2(-\infty, \infty).$$

Suppose that the function $f(s)$ satisfies the condition

$$f(s) \in W_2^1(-\infty, \infty).$$

Applying the Fourier transform to equation (2.2.13), we obtain

$$\bar{f}(\omega)\bar{K}(\omega) = \bar{\psi}(\omega), \quad (2.2.15)$$

where $\bar{K}(\omega) = (1 + |\omega| b)\mathrm{e}^{-|\omega| b}$, $\bar{f}(\omega)$, and $\bar{\psi}(\omega)$ are the Fourier images of the functions $K(y)$, $f(s)$, and $\psi(x)$.

Evidently, $\bar{K}(\omega) \neq 0$ almost everywhere. This provides the uniqueness of solution of equation (2.2.13).

In practice, we usually know some δ-approximation of the function $\psi(x)$. Namely, we know $\psi_\delta(x)$ such that

$$\|\psi(x) - \psi_\delta(x)\|_{L_2} \le \delta.$$

Consider the Tikhonov functional (see Tikhonov and Arsenin, 1977)

$$M^\alpha[f, \psi_\delta] = \|A[x, f(s)] - \psi_\delta(x)\|_{L_2}^2 + \alpha \|f(s)\|_{W_2^1}^2, \quad (2.2.16)$$

$$M^{\alpha}[f,\psi] = \|A[x,f(s)] - \psi(x)\|^2_{L_2} + \alpha\|f(s)\|^2_{W_2^1}$$
$$= \int_{-\infty}^{\infty}\Big[\int_{-\infty}^{\infty} K(x-s)f(s)\,\mathrm{d}s - \psi(x)\Big]^2\,\mathrm{d}x$$
$$+\,\alpha\int_{-\infty}^{\infty}\Big\{\Big[\frac{\mathrm{d}f(x)}{\mathrm{d}x}\Big]^2 + f^2(x)\Big\}\,\mathrm{d}x. \qquad (2.2.17)$$

In Tikhonov and Arsenin (1977), it is shown that for each $\alpha > 0$ there exists a unique element $f_\delta^\alpha(s)$ realizing the minimum of the functional (2.2.16). In this case, if $\alpha(\delta) \to 0$ so that $\delta^2/\alpha(\delta) \to 0$, then the sequence $f_\delta^{\alpha(\delta)}(s)$ converges to $f(s)$, the exact solution of (2.2.13), by the norm W_2^1. From the imbedding theorem (Sobolev, 1950) it follows that the convergence of approximations to the exact solution is uniform. Besides, there exists a unique function $f^\alpha(s)$ which realizes the minimum of the functional $M^\alpha[f,\psi]$ for a fixed $\alpha > 0$. In this case $f^\alpha(s) \to f(s)$ for $\alpha \to 0$ (Tikhonov and Arsenin, 1977).

Varying the functional $M^\alpha[f,\psi]$, we obtain the Euler equation

$$\alpha\Big[\frac{\mathrm{d}^2 f(x)}{\mathrm{d}x^2} - f(x)\Big] = \int_{-\infty}^{\infty} K_1(x-s)f(s)\,\mathrm{d}s - \int_{-\infty}^{\infty} K_1(x-t)\psi(t)\,\mathrm{d}t,$$

where

$$K_1(x-s) = \int_{-\infty}^{\infty} K(x-t)K(t-s)\,\mathrm{d}t.$$

Applying the Fourier transform to the Euler equation, we find the function $f^\alpha(s)$ which minimizes the functional $M^\alpha[f,\psi]$

$$f^\alpha(s) = \frac{1}{2\pi}\int_{-\infty}^{\infty}\frac{\bar{K}(\omega)\bar{\psi}(\omega)\mathrm{e}^{\mathrm{i}\omega s}}{L(\omega)+\alpha(\omega^2+1)}\,\mathrm{d}\omega. \qquad (2.2.18)$$

Analogously, we determine the function $f_\delta^\alpha(s)$ which minimizes the functional $M^\alpha[f,\psi_\delta]$

$$f_\delta^\alpha(s) = \frac{1}{2\pi}\int_{-\infty}^{\infty}\frac{\bar{K}(\omega)\bar{\psi}_\delta(\omega)\mathrm{e}^{\mathrm{i}\omega s}}{L(\omega)+\alpha(\omega^2+1)}\,\mathrm{d}\omega. \qquad (2.2.19)$$

Here $L(\omega) = \bar{K}^2(\omega)$; $\bar{K}(\omega)$, $\bar{\psi}(\omega)$, and $\bar{\psi}_\delta(\omega)$ are the Fourier images of the functions $K(y)$, $\psi(x)$, and $\psi_\delta(x)$.

Now we obtain the accuracy estimates of the approximation $f_\delta^\alpha(s)$ to the exact solution $f(s)$ in the norm of the spaces C and L_2.

Evidently,

$$\|f(s) - f_\delta^\alpha(s)\| \le \|f(s) - f^\alpha(s)\| + \|f^\alpha(s) - f_\delta^\alpha(s)\|. \tag{2.2.20}$$

Formulas (2.2.15), (2.2.18), and (2.2.19) yield

$$f(s) - f^\alpha(s) = \frac{\alpha}{2\pi}\int_{-\infty}^{\infty} \frac{(\omega^2+1)\bar f(\omega)\mathrm{e}^{\mathrm{i}\omega s}}{L(\omega) + \alpha(\omega^2+1)}\,\mathrm{d}\omega, \tag{2.2.21}$$

$$f^\alpha(s) - f_\delta^\alpha(s) = \frac{\alpha}{2\pi}\int_{-\infty}^{\infty} \frac{\bar K(\omega)[\bar\psi(\omega) - \bar\psi_\delta(\omega)]\mathrm{e}^{\mathrm{i}\omega s}}{L(\omega) + \alpha(\omega^2+1)}\,\mathrm{d}\omega. \tag{2.2.22}$$

It remains to estimate (2.2.21) and (2.2.22) as functions of α and δ. Similar to Koshlyakov *et al.* (1970), we obtain

$$\|f^\alpha(s) - f_\delta^\alpha(s)\|_{L_2} \le \frac{\delta}{\sqrt{\alpha}}, \tag{2.2.23}$$

$$\|f^\alpha(s) - f_\delta^\alpha(s)\|_C \le \frac{\delta}{\sqrt{2\alpha}}. \tag{2.2.24}$$

Now we estimate (2.2.21) in the metric of the space C. Since the function $|\bar f(\omega)|$ is even, therefore, from (2.2.21) we obtain

$$\|f(s) - f^\alpha(s)\|_C \le \frac{\alpha}{\pi}\int_0^\infty \frac{(\omega^2+1)|\bar f(\omega)|}{L(\omega) + \alpha(\omega^2+1)}\,\mathrm{d}\omega.$$

Write the last integral in the form

$$\begin{aligned} J(\alpha) &= \frac{\alpha}{\pi}\int_0^N \frac{(\omega^2+1)|\bar f(\omega)|}{L(\omega) + \alpha(\omega^2+1)}\,\mathrm{d}\omega + \frac{\alpha}{\pi}\int_N^\infty \frac{(\omega^2+1)|\bar f(\omega)|}{L(\omega) + \alpha(\omega^2+1)}\,\mathrm{d}\omega \\ &= J_1(\alpha) + J_2(\alpha), \end{aligned} \tag{2.2.25}$$

where N is a positive number.

It is easy to obtain a lower bound for $J(\alpha)$

$$J(\alpha) \ge \mathrm{const}\cdot\alpha. \tag{2.2.26}$$

Now we obtain an upper bound. We have

$$\begin{aligned} J_2(\alpha) &\le \frac{1}{\pi}\int_N^\infty |\bar f(\omega)|\,\mathrm{d}\omega \le \frac{1}{\pi}\Big[\int_N^\infty \omega^2|\bar f(\omega)|^2\,\mathrm{d}\omega\Big]^{1/2}\Big[\int_N^\infty \frac{\mathrm{d}\omega}{\omega^2}\Big]^{1/2} \\ &\le \frac{A_1}{\sqrt{N}}, \end{aligned} \tag{2.2.27}$$

where

$$A_1 = \frac{1}{\pi}\Big[\int_N^\infty \omega^2|\bar{f}(\omega)|^2\,d\omega\Big]^{1/2} < \infty, \qquad \text{because} \quad f(s) \in W_2^1.$$

Now we obtain the estimate for $J_1(\alpha)$

$$\begin{aligned} J_1(\alpha) &= \frac{\alpha}{\pi}\int_0^N \frac{(\omega^2+1)|\bar{f}(\omega)|}{L(\omega)+\alpha(\omega^2+1)}\,d\omega \le \frac{\alpha}{\pi}\int_{-\infty}^\infty \frac{(\omega^2+1)e^{i\omega s}}{(1+\omega b)^2}|\bar{f}(\omega)|\,d\omega \\ &\le \frac{\alpha}{\pi}\Big[\int_0^N \frac{(\omega^2+1)^2 e^{4\omega s}}{(1+\omega b)^4}\,d\omega\Big]^{1/2}\Big[\int_0^N |\bar{f}(\omega)|^2\,d\omega\Big]^{1/2} \\ &\le \alpha\frac{\sqrt{N}(N^2+1)e^{2Nb}}{(1+Nb)^2}A_2, \end{aligned} \tag{2.2.28}$$

where

$$A_2 = \frac{1}{\pi}\Big[\int_0^\infty |\bar{f}(\omega)|^2\,d\omega\Big]^{1/2} < \infty,$$

because $f(s) \in W_2^1$.

The estimates (2.2.27) and (2.2.28) yield

$$\begin{aligned} J(\alpha) &\le \alpha\frac{\sqrt{N}(N^2+1)e^{2Nb}}{(1+Nb)^2}A_2 + \frac{A_1}{\sqrt{N}} \\ &= \frac{A_1}{\sqrt{N}}\Big[\alpha\frac{N(N^2+1)e^{2Nb}A_2}{(1+Nb)^2 A_1} + 1\Big] \le \frac{A_1}{\sqrt{N}}\big[\alpha B_1 e^{3Nb} + 1\big], \end{aligned}$$

where

$$B_1 = \frac{A_2}{A_1}\max_N\Big[\frac{N(N^2+1)e^{-Nb}}{(1+Nb)^2}\Big].$$

We find N from the equation

$$\alpha B_1 e^{3Nb} = 1,$$

whence

$$N = \frac{1}{3b}\ln\frac{1}{\alpha B_1}.$$

Therefore,

$$J(\alpha) \le 2A_1\Big[\frac{3b}{\ln(\alpha B_1)^{-1}}\Big]^{1/2}. \tag{2.2.29}$$

Now formulas (2.2.20), (2.2.24), and (2.2.29) yield

$$\|f(s) - f_\delta^\alpha(s)\|_C \le 2A_1\Big[\frac{3b}{\ln(\alpha B_1)^{-1}}\Big]^{1/2} + \frac{\delta}{\sqrt{2\alpha}}$$

$$= 2A_1\Big[\frac{3b}{\ln(\alpha B_1)^{-1}}\Big]^{1/2}\Big[1+\delta\frac{1}{2A_1}\Big(\frac{\ln(\alpha B_1)^{-1}}{6b\alpha}\Big)^{1/2}\Big]$$
$$\le 2A_1\Big[\frac{3b}{\ln(\alpha B_1)^{-1}}\Big]^{1/2}\Big[1+\frac{1}{2A_1\sqrt{6bB_1}}\frac{\delta}{\alpha}\Big].$$

The regularization parameter α can be found from the equation

$$\frac{1}{2A_1\sqrt{6bB_1}}\frac{\delta}{\alpha} = 1,$$

whence

$$\alpha = \frac{\delta}{2A_1\sqrt{6bB_1}}.$$

Therefore,

$$\|f(s) - f_\delta^\alpha(s)\|_C \le 4A_1\Big[\frac{3b}{\ln((2A_1/\delta)\sqrt{6b/B_1})}\Big]^{1/2}. \tag{2.2.30}$$

We now obtain another estimate for $f(s) - f_\delta^\alpha(s)$. From (2.2.27) and (2.2.28) we have

$$J(\alpha) \le \alpha\frac{\sqrt{N}(N^2+1)e^{2Nb}}{(1+Nb)^2}A_2 + \frac{A_1}{\sqrt{N}} \le \alpha C_1 e^{3Nb} + \frac{A_1}{\sqrt{N}}, \tag{2.2.31}$$

where

$$C_1 = A_2\max_N\Big\{\frac{\sqrt{N}(N^2+1)e^{-Nb}}{(1+Nb)^2}\Big\}.$$

Minimizing the right-hand side of (2.2.31) with respect to N, we obtain

$$J(\alpha) \le \alpha C_1 e^{3N(\alpha)b} + \frac{A_1}{\sqrt{N(\alpha)}}. \tag{2.2.32}$$

Here $N(\alpha)$ is the root of the equation

$$3bN + \frac{3}{2}\ln N = \ln\Big(\frac{A_1}{6C_1\alpha b}\Big). \tag{2.2.33}$$

Thus, from (2.2.20), (2.2.24), and (2.2.32) we obtain

$$\|f(s) - f_\delta^\alpha(s)\|_C \le \alpha C_1 e^{3bN(\alpha)} + \frac{A_1}{\sqrt{N(\alpha)}} + \frac{\delta}{\sqrt{2\alpha}}. \tag{2.2.34}$$

Minimizing the right-hand side of (2.2.34) with respect to α, we obtain

$$\|f(s) - f_\delta^\alpha(s)\|_C \le \alpha(\delta) C_1 \mathrm{e}^{3bN(\alpha(\delta))} + \frac{A_1}{\sqrt{N(\alpha(\delta))}} + \frac{\delta}{\sqrt{2\alpha(\delta)}}, \qquad (2.2.35)$$

where $\alpha(\delta)$ is the positive root of the equation

$$[1 + 3b\alpha N'(\alpha)]\alpha^{3/2} C_1 \mathrm{e}^{3bN(\alpha)} - \frac{A_1}{2}\Big[\frac{\alpha}{N(\alpha)}\Big]^{3/2} N'(\alpha) = \frac{\delta}{2\sqrt{2}}. \qquad (2.2.36)$$

From (2.2.33) and (2.2.36) it follows that $\alpha(\delta) \to 0$ as $\delta \to 0$. Now let us estimate (2.2.21) in the L_2-metric. From (2.2.21), by the Plancherel theorem (Bochner, 1959), we obtain

$$\|f(s) - f^\alpha(s)\|_{L_2}^2 \le \frac{\alpha^2}{\pi}\int_0^\infty \frac{(\omega^2+1)^2|\bar{f}(\omega)|^2}{[L(\omega) + \alpha(\omega^2+1)]^2}\,\mathrm{d}\omega.$$

Let us denote the last integral by $\bar{J}(\alpha)$ and rewrite it in the form

$$\bar{J}(\alpha) = \frac{\alpha^2}{\pi}\int_0^M \frac{(\omega^2+1)^2|\bar{f}(\omega)|^2}{[L(\omega) + \alpha(\omega^2+1)]^2}\,\mathrm{d}\omega + \frac{\alpha^2}{\pi}\int_M^\infty \frac{(\omega^2+1)^2|\bar{f}(\omega)|^2}{[L(\omega) + \alpha(\omega^2+1)]^2}\,\mathrm{d}\omega$$
$$= \bar{J}_1(\alpha) + \bar{J}_2(\alpha),$$

where M is a positive number. Analogously to (2.2.26) we find

$$\bar{J}(\alpha) \ge \mathrm{const}\cdot\alpha^2. \qquad (2.2.37)$$

Now estimate $\bar{J}_2(\alpha)$:

$$\bar{J}_2(\alpha) \le \frac{1}{\pi}\int_M^\infty |\bar{f}(\omega)|^2\,\mathrm{d}\omega = \frac{1}{\pi}\int_M^\infty \frac{1}{\omega^2}\omega^2|\bar{f}(\omega)|^2\,\mathrm{d}\omega$$
$$\le \frac{1}{\pi M^2}\int_M^\infty \omega^2|\bar{f}(\omega)|^2\,\mathrm{d}\omega \le \frac{\bar{A}_1^2}{M^2}, \qquad (2.2.38)$$

where

$$\bar{A}_1 = \Big\{\frac{1}{\pi}\int_0^\infty \omega^2|\bar{f}(\omega)|^2\,\mathrm{d}\omega\Big\}^{1/2} < \infty.$$

We now estimate $\bar{J}_1(\alpha)$. To this end, we require in addition that

$$|\bar{f}(\omega)| \le d, \qquad d = \mathrm{const} > 0.$$

Then

$$\bar{J}_1(\alpha) = \frac{\alpha^2}{\pi}\int_0^M \frac{(\omega^2+1)^2|\bar{f}(\omega)|^2}{[L(\omega) + \alpha(\omega^2+1)]^2}\,\mathrm{d}\omega + \frac{\alpha^2}{\pi}\int_M^\infty \frac{(\omega^2+1)^2\mathrm{e}^{4\omega b}|\bar{f}(\omega)|^2}{(1+\omega b)^4}\,\mathrm{d}\omega$$

$$\leq \alpha^2 \frac{M(M^2+1)^2 e^{4Mb}}{\pi(1+Mb)^4} d^2. \tag{2.2.39}$$

The estimates (2.2.38) and (2.2.39) yield

$$\begin{aligned} \bar{J}(\alpha) &\leq \alpha^2 \frac{M(M^2+1)^2 e^{4Mb}}{\pi(1+Mb)^4} d^2 + \frac{\bar{A}_1^2}{M^2} \\ &= \frac{\bar{A}_1^2}{M^2}\Big[\alpha^2 \frac{M^3(M^2+1)^2 e^{4Mb}}{\pi(1+Mb)^4 \bar{A}_1^2} d^2 + 1\Big] \leq \frac{\bar{A}_1^2}{M^2}(\alpha^2 B_2 e^{5Mb} + 1), \end{aligned}$$

where

$$B_2 = \frac{d^2}{\pi \bar{A}_1^2} \max_M \Big\{ \frac{M^3(M^2+1)^2 e^{-Mb}}{(1+Mb)^4} \Big\}.$$

From the equation

$$\alpha^2 B_2 e^{5Mb} = 1$$

we find

$$M = \frac{2}{5b} \ln \frac{1}{\alpha\sqrt{B_2}}.$$

Therefore,

$$\bar{J}(\alpha) \leq \Big[\frac{5\bar{A}_1 b}{2\ln(\alpha\sqrt{B_2})^{-1}} \Big]^2. \tag{2.2.40}$$

Formulas (2.2.20), (2.2.23), and (2.2.40) yield

$$\begin{aligned} \|f(s) - f_\delta^\alpha(s)\|_{L_2} &\leq \frac{5\bar{A}_1 b}{2\ln(\alpha\sqrt{B_2})^{-1}} + \frac{\delta}{\sqrt{\alpha}} \\ &= \frac{5\bar{A}_1 b}{2\ln(\alpha\sqrt{B_2})^{-1}} \Big[1 + \frac{2\ln(\alpha\sqrt{B_2})^{-1}}{5\bar{A}_1 b} \frac{\delta}{\sqrt{\alpha}} \Big] \\ &\leq \frac{5\bar{A}_1 b}{2\ln(\alpha\sqrt{B_2})^{-1}} \Big[1 + \frac{2}{5\bar{A}_1 b\sqrt{B_2}} \frac{\delta}{\alpha^{3/2}} \Big]. \end{aligned} \tag{2.2.41}$$

From the equation

$$\frac{2}{5\bar{A}_1 b\sqrt{B_2}} \frac{\delta}{\alpha^{3/2}} = 1$$

we find the regularization parameter α:

$$\alpha = \Big[\frac{2\delta}{5\bar{A}_1 b\sqrt{B_2}} \Big]^{2/3}.$$

Hence, by (2.2.41),

$$\|f(s) - f_\delta^\alpha(s)\|_{L_2} \le \frac{15\bar{A}_1 b}{2\ln\left[5A_1 b/(2B_2^{1/4}\delta)\right]}. \tag{2.2.42}$$

Let us find another estimate for $f(s) - f_\delta^\alpha(s)$ in the metric of the space L_2. From (2.2.38) and (2.2.39) we obtain

$$\bar{J}(\alpha) \le \alpha^2 \frac{M(M^2+1)^2 e^{4Mb}}{\pi(1+Mb)^4} d^2 + \frac{\bar{A}_1^2}{M^2} \le \alpha^2 C_2 e^{5Mb} + \frac{\bar{A}_1^2}{M^2}, \tag{2.2.43}$$

where

$$C_2 = \frac{d^2}{\pi} \max_M \left\{\frac{M(M^2+1)^2 e^{-Mb}}{(1+Mb)^4}\right\}.$$

Minimizing the right-hand side of (2.2.43) with respect to M, we obtain

$$\bar{J}(\alpha) \le \alpha^2 C_2 e^{5M(\alpha)b} + \frac{\bar{A}_1^2}{M^2(\alpha)}, \tag{2.2.44}$$

where $M(\alpha)$ is the root of the equation

$$5bM + 3\ln M = 2\ln\left(\frac{\bar{A}_1}{\alpha}\sqrt{\frac{2}{5bC_2}}\right). \tag{2.2.45}$$

From (2.2.44), (2.2.20), and (2.2.23) we get

$$\begin{aligned} \|f(s) - f_\delta^\alpha(s)\|_{L_2}^2 &\le \left\{\left[\alpha^2 C_2 e^{5M(\alpha)b} + \frac{\bar{A}_1^2}{M^2(\alpha)}\right]^{1/2} + \frac{\delta}{\sqrt{\alpha}}\right\}^2 \\ &\le 2\alpha^2 C_2 e^{5M(\alpha)b} + \frac{2\bar{A}_1^2}{M^2(\alpha)} + \frac{2\delta^2}{\alpha}. \end{aligned} \tag{2.2.46}$$

Minimizing the right-hand side of (2.2.46) with respect to α, we obtain

$$\|f(s) - f_\delta^\alpha(s)\|_{L_2} \le \sqrt{2}\left\{\alpha^2(\delta) C_2 e^{5M(\alpha(\delta))b} + \frac{\bar{A}_1^2}{M^2(\alpha(\delta))} + \frac{\delta^2}{\alpha}\right\}^{1/2}, \tag{2.2.47}$$

where $\alpha(\delta)$ is the positive root of the equation

$$[2 + 5b\alpha M'(\alpha)]\alpha^3 C_2 e^{5bM(\alpha)} - \frac{2\alpha^2 \bar{A}_1^2 M'(\alpha)}{M^3(\alpha)} = \delta^2. \tag{2.2.48}$$

Relations (2.2.45) and (2.2.48) yield $\alpha(\delta) \to 0$ as $\delta \to 0$.

We have presented several equations from which the regularization parameter α is obtained. Some equations for determining α are also considered in Tikhonov and Arsenin (1977).

Suppose that $\varphi_1(x)$ and $\varphi_2(x)$ are known approximately; i. e., instead of them, we know their approximations $\varphi_{1\varepsilon}(x)$ and $\varphi_{2\varepsilon}(x)$ such that

$$|\varphi_j(x) - \varphi_{j\varepsilon}(x)| \le h(x,\varepsilon), \qquad j = 1,2, \tag{2.2.49}$$

where $0 \le h(x,\varepsilon) \in L_1 \cap L_2$. Here $h(x,\varepsilon) \to 0$ as $\varepsilon \to 0$. Therefore, instead of $\psi(x)$, we have the function $\psi_\delta(x)$:

$$\|\psi(x) - \psi_\delta(x)\|_{L_2} \le a(\varepsilon) = \delta. \tag{2.2.50}$$

Here

$$\begin{aligned} a(\varepsilon) &= \Big[\int_{-\infty}^{\infty} h^2(x,\varepsilon)\,\mathrm{d}x\Big]^{1/2} \\ &+ \Big[\int_{-\infty}^{\infty} h^2(x,\varepsilon)\,\mathrm{d}x\Big]^{1/2}\Big[\frac{b}{\pi^2}\int_{-\infty}^{\infty}\int_{-\infty}^{\infty}\frac{h(tb+x,\varepsilon)}{(1+t^2)^2}\,\mathrm{d}t\,\mathrm{d}x\Big]^{1/2}. \end{aligned}$$

As an approximate solution, we take the function

$$u_\varepsilon^\alpha(x,y) = \frac{2}{\pi}\int_{-\infty}^{\infty}\frac{y^3 f_\varepsilon^\alpha(s)\,\mathrm{d}s}{[(x-s)^2+y^2]^2} + \frac{1}{\pi}\int_{-\infty}^{\infty}\frac{y^2\varphi_{1\varepsilon}(s)\,\mathrm{d}s}{(x-s)^2+y^2}, \tag{2.2.51}$$

where $f_\varepsilon^\alpha(s)$ is the approximate solution of the integral equation (2.2.13).

Now estimate $u(x,y) - u_\varepsilon^\alpha(x,y)$ in the metric of the space C.

Formulas (2.2.12), (2.2.30), and (2.2.49)–(2.2.51) yield

$$\|u(x,y) - u_\varepsilon^\alpha(x,y)\|_C \le 4A_1\left[\frac{3b}{\ln\big[(A_1/a(\varepsilon))\sqrt{6b/B_1}\,\big]}\right]^{1/2} + \frac{a_1(\varepsilon)}{\pi}, \tag{2.2.52}$$

where

$$a_1(\varepsilon) = \int_{-\infty}^{\infty} h(x,\varepsilon)\,\mathrm{d}x.$$

Relations (2.2.12), (2.2.35), and (2.2.49)–(2.2.51) imply

$$\|u(x,y) - u_\varepsilon^\alpha(x,y)\|_C \le \alpha(\varepsilon)C_1\mathrm{e}^{3bN(\alpha(\varepsilon))} + \frac{A_1}{\sqrt{N(\alpha(\varepsilon))}} + \frac{a(\varepsilon)}{\sqrt{2\alpha(\varepsilon)}} + \frac{a_1(\varepsilon)}{\pi}, \tag{2.2.53}$$

where $N(\alpha(\varepsilon))$ and $\alpha(\varepsilon)$ are the positive roots of equations (2.2.33) and (2.2.36) for $\delta = a(\varepsilon)$.

Now we estimate $u(x,y) - u_\varepsilon^\alpha(x,y)$ in the metric of the space L_2:

$$\begin{aligned} u(x,y) - u_\varepsilon^\alpha(x,y) &= \int_{-\infty}^{\infty} K_2(x-s,y)[f(s) - f_\varepsilon^\alpha(s)]\,\mathrm{d}s \\ &\quad + \int_{-\infty}^{\infty} K_3(x-s,y)[\varphi_1(s) - \varphi_{1\varepsilon}(s)]\,\mathrm{d}s \\ &= F_1(x,y) + F_2(x,y), \end{aligned} \tag{2.2.54}$$

where

$$K_2(x-s,y) = \frac{2}{\pi} y^3 [(x-s)^2 + y^2]^{-2},$$

$$K_3(x-s,y) = \frac{2}{\pi} y^2 [(x-s)^2 + y^2]^{-1}.$$

Taking into account the Plancherel theorem (Bochner, 1959), we obtain

$$\begin{aligned} \|F_1(x,y)\|_{L_2} &= \frac{1}{\sqrt{2\pi}} \|\bar{F}_1(\omega,y)\|_{L_2} = \frac{1}{\sqrt{2\pi}} \|\bar{K}_2(\omega,y)[\bar{f}(\omega) - \bar{f}_\varepsilon^\alpha(\omega)]\|_{L_2} \\ &\le \frac{1}{\sqrt{2\pi}} \|\bar{f}(\omega) - \bar{f}_\varepsilon^\alpha(\omega)\|_{L_2} \le \|\bar{f}(\omega) - \bar{f}_\varepsilon^\alpha(\omega)\|_{L_2}, \end{aligned} \tag{2.2.55}$$

where

$$\bar{F}_1(\omega,y), \quad \bar{K}_2(\omega,y) = (1 + |\omega|y)\mathrm{e}^{-|\omega|y} \le 1, \quad \bar{f}(\omega), \quad \bar{f}_\varepsilon^\alpha(\omega)$$

are the Fourier images of the functions

$$F_1(x,y), \quad K_2(x,y), \quad f(s), \quad f_\varepsilon^\alpha(s).$$

By (2.2.49) and the Cauchy – Schwarz – Bunyakovskii inequality, we have

$$\|F_2(x,y)\|_{L_2} \le a_2(x,\varepsilon), \tag{2.2.56}$$

where

$$a_2(y,\varepsilon) = \Big[\int_{-\infty}^{\infty} h(x,\varepsilon)\,\mathrm{d}x\Big]^{1/2} \Big[\int_{-\infty}^{\infty}\int_{-\infty}^{\infty} \frac{yh(ty+x,\varepsilon)}{\pi^2(1+t^2)^2}\,\mathrm{d}t\,\mathrm{d}x\Big]^{1/2}.$$

Formulas (2.2.42), (2.2.49), (2.2.50), and (2.2.54)–(2.2.56) yield

$$\|u(x,y) - u_\varepsilon^\alpha(x,y)\|_{L_2} \le \frac{15\bar{A}_1 b}{2\ln[5\bar{A}_1 b/(2B_2^{1/4} a(\varepsilon))]}. \tag{2.2.57}$$

In virtue of (2.2.47), (2.2.49), (2.2.50), and (2.2.54)–(2.2.56), we have

$$\|u(x,y)-u_\varepsilon^\alpha(x,y)\|_{L_2} \leq \sqrt{2}\left[\alpha^2(\varepsilon)C_2 e^{5bM(\alpha(\varepsilon))}+\frac{\bar{A}_1^2}{M^2(\alpha(\varepsilon))}+\frac{a^2(\varepsilon)}{\alpha(\varepsilon)}\right]^{1/2}+a_2(y,\varepsilon). \quad (2.2.58)$$

Here $M(\alpha(\varepsilon))$ and $\alpha(\varepsilon)$ are the positive roots of equations (2.2.45) and (2.2.48) for $\delta = a(\varepsilon)$.

We assume that

$$\varphi_1(x)=\tilde{\varphi}_1(x)\beta(x), \quad |\tilde{\varphi}_1(x)|\leq M_1, \quad 0\leq \beta(x)\in L_1\cap L_2.$$

We take the function

$$u_\varepsilon^{\alpha i}(x,y)=\frac{2}{\pi}\int_{-i}^{i}\frac{f_\varepsilon^\alpha(x+ty)}{(1+t^2)^2}\,dt+\frac{1}{\pi}\int_{-i}^{i}\frac{y\varphi_{1\varepsilon}(x+ty)}{1+t^2}\,dt \quad (2.2.59)$$

as an approximate solution of the problem (2.2.1)–(2.2.4). Here $f_\varepsilon^\alpha(s)$ is the approximate solution of the integral equation (2.2.13).

Evidently,

$$|u(x,y)-u_\varepsilon^{\alpha i}(x,y)|\leq |u(x,y)-u_\varepsilon^\alpha(x,y)|+|u_\varepsilon^\alpha(x,y)-u_\varepsilon^{\alpha i}(x,y)|, \quad (2.2.60)$$

$$|f_\varepsilon^\alpha(s)|\leq |f(s)-f_\delta^\alpha(s)|+|f(s)|, \quad (2.2.61)$$

$$|\varphi_{1\varepsilon}(s)|\leq |\varphi_1(s)-\varphi_{1\varepsilon}(s)|+|\varphi_1(s)|. \quad (2.2.62)$$

Formulas (2.2.12), (2.2.49)–(2.2.52), and (2.2.59)–(2.2.62) yield

$$\|u(x,y)-u_\varepsilon^{\alpha i}(x,y)\|_C \leq 8A_1\left[\frac{3b}{\ln\left[(A_1/a(\varepsilon))\sqrt{6b/B_1}\right]}\right]^{1/2}+\frac{2a_1(\varepsilon)}{\pi}+\frac{4M_2}{3\pi i^2}+\frac{M_1}{\pi}a_3(y,i),$$

where

$$M_1=\max|\tilde{\varphi}_1(x)|, \qquad M_2=\max|f(x)|,$$

$$a_3(y,i)=\max\left[\int_i^\infty \frac{y[\beta(x-ty)+\beta(x+ty)]}{1+t^2}\,dt\right].$$

The number i is found from the equation

$$\frac{1}{i^3}=a_1(\varepsilon).$$

Thus,

$$i(\varepsilon) = \left[\frac{1}{a_1(\varepsilon)}\right]^{1/3}. \tag{2.2.63}$$

Therefore, we have

$$\|u(x,y) - u_\varepsilon^{\alpha i}(x,y)\|_C$$
$$\leq 8A_1\left[\frac{3b}{\ln\left[(A_1/a(\varepsilon))\sqrt{6b/B_1}\,\right]}\right]^{1/2} + \left(2 + \frac{4M_2}{3}\right)\frac{2a_1(\varepsilon)}{\pi} + \frac{M_1}{\pi}a_3(y,i).$$

From (2.2.12), (2.2.49)–(2.2.51), (2.2.53), and (2.2.59)–(2.2.63) we obtain

$$\|u(x,y) - u_\varepsilon^{\alpha i}(x,y)\|_C \leq 2\alpha(\varepsilon)C_1\,\mathrm{e}^{3bN(\alpha(\varepsilon))} + \frac{2A_1}{\sqrt{N(\alpha(\varepsilon))}} + \frac{2a(\varepsilon)}{\sqrt{2\alpha(\varepsilon)}}$$
$$+ \left(2 + \frac{2M_2}{3}\right)\frac{a_1(\varepsilon)}{\pi} + \frac{M_1}{\pi}a_3(y, i(\varepsilon)).$$

Here $N(\alpha(\varepsilon))$ and $\alpha(\varepsilon)$ are the positive roots of equations (2.2.33) and (2.2.36) for $\delta = a(\varepsilon)$.

Let us estimate $u(x,y) - u_\varepsilon^{\alpha i}(x,y)$ in the metric of the space L_2.

Formulas (2.2.12), (2.2.49)–(2.2.51), (2.2.57), and (2.2.59)–(2.2.61) yield

$$\|u(x,y) - u_\varepsilon^{\alpha i}(x,y)\|_{L_2} \leq \frac{15\bar{A}_1 b}{\ln\left[5\bar{A}_1 b/(2B_2^{1/4}a(\varepsilon))\right]} + 2a_2(y,\varepsilon)$$
$$+ \frac{M_1\bar{C}_1(y,i)}{\sqrt{3}\,\pi}\cdot\frac{1}{i^{3/2}} + \frac{2\bar{C}_2(y,i)}{\sqrt{7}\,\pi}\cdot\frac{1}{i^{7/2}}, \tag{2.2.64}$$

where

$$\bar{C}_1(y,i) = \left[\int_{-\infty}^{\infty}\int_i^{\infty} y^2\beta^2(x-ty)\,\mathrm{d}t\,\mathrm{d}x\right]^{1/2}$$
$$+ \left[\int_{-\infty}^{\infty}\int_i^{\infty} y^2\beta^2(x+ty)\,\mathrm{d}t\,\mathrm{d}x\right]^{1/2},$$
$$\bar{C}_2(y,i) = \left[\int_{-\infty}^{\infty}\int_i^{\infty} f^2(x-ty)\,\mathrm{d}t\,\mathrm{d}x\right]^{1/2} + \left[\int_{-\infty}^{\infty}\int_i^{\infty} f^2(x+ty)\,\mathrm{d}t\,\mathrm{d}x\right]^{1/2}.$$

Here the number i satisfies the equation

$$\frac{1}{i^{3/2}} = a(\varepsilon),$$

whence

$$i(\varepsilon) = \left[\frac{1}{a(\varepsilon)}\right]^{2/3}. \tag{2.2.65}$$

Substituting $i(\varepsilon)$ into (2.2.64), we obtain

$$\|u(x,y) - u_\varepsilon^{\alpha i}(x,y)\|_{L_2} \le \frac{15\bar{A}_1 b}{\ln\left[5\bar{A}_1 b/(2B_2^{1/4}a(\varepsilon))\right]} + 2a_2(y,\varepsilon) + \frac{M_1 a(\varepsilon)\bar{C}_1(y,i(\varepsilon))}{\sqrt{3}\,\pi} + \frac{2[a(\varepsilon)]^{7/3}\bar{C}_2(y,i(\varepsilon))}{\sqrt{7}\,\pi}.$$

From formulas (2.2.12), (2.2.49)–(2.2.51), (2.2.58)–(2.2.62), and (2.2.65) we have

$$\|u(x,y) - u_\varepsilon^{\alpha i}(x,y)\|_{L_2} \le 2\sqrt{2}\left[\alpha^2(\varepsilon)C_2 e^{5bM(\alpha(\varepsilon))} + \frac{\bar{A}_1^2}{M^2(\alpha(\varepsilon))} + \frac{a^2(\varepsilon)}{\alpha(\varepsilon)}\right]^{1/2} + 2a_2(y,\varepsilon) + \frac{M_1 a(\varepsilon)}{\sqrt{3}\,\pi}\bar{C}_1(y,i(\varepsilon)) + \frac{2[a(\varepsilon)]^{7/3}}{\sqrt{7}\,\pi}\bar{C}_2(y,i(\varepsilon)).$$

Here $M(\alpha(\varepsilon))$ and $\alpha(\varepsilon)$ are the positive roots of equations (2.2.45) and (2.2.48) for $\delta = \alpha(\varepsilon)$.

2.3. THE SECOND INTERNAL BOUNDARY VALUE PROBLEM IN A RECTANGLE

Case A. Suppose that D_1, D_2 ($D_1 \supset \overline{D_2}$) are the following rectangles

$$D_1 = \{(x,y) \mid 0 < x < a_3,\ 0 < y < b_3\},$$
$$D_2 = \{(x,y) \mid a_1 < x < a_2,\ b_1 < y < b_2\}.$$

Let $S_1 = \partial D_1$, $S_2 = \partial D_2$; and let ν be the external normal to S_1.

Problem 2.3.1. We need to find the function $u(x,y)$ in D_1 satisfying the conditions

$$\Delta^2 u(x,y) = 0 \quad \text{in} \quad D_1, \tag{2.3.1}$$

$$u'_x(0,y) = \psi_1(y), \qquad u'_x(a_3,y) = \psi_2(y), \qquad 0 \le y \le b_3, \tag{2.3.2}$$

$$u'_y(x,0) = \varphi_1(x), \qquad u'_y(x,b_3) = \varphi_2(x), \qquad 0 \le x \le a_3, \tag{2.3.3}$$

$$u(a_1,y) = \psi_3(y), \qquad u(a_2,y) = \psi_4(y), \qquad b_1 \le y \le b_2, \tag{2.3.4}$$

$$u(x, b_1) = \varphi_3(x), \qquad u(x, b_2) = \varphi_4(x), \qquad a_1 \le x \le a_3. \tag{2.3.5}$$

Here, as before, Δ is the Laplace operator; $\psi_j(y)$ and $\varphi_j(x)$, $j = 1, 2, 3, 4$, are given functions such that $\psi_3(b_1) = \varphi_3(a_1)$, $\psi_3(b_2) = \varphi_4(a_1)$, $\psi_4(b_1) = \varphi_3(a_2)$, and $\psi_4(b_2) = \varphi_4(a_2)$.

Without any loss of generality, we may assume that

$$\psi_j(b_1) = \psi_j(b_2) = \varphi_j(a_1) = \varphi_j(a_2) = 0, \qquad j = 3, 4.$$

Indeed, if it is not so, we subtract the biharmonic function

$$u_0(x, y) = A + Bx + Cy + Dxy$$

from the function $u(x, y)$. Here

$$A = \frac{[\varphi_3(a_1)a_2 - \varphi_3(a_2)a_1]b_2 - [\varphi_4(a_1)a_2 - \varphi_4(a_2)a_1]b_1}{(a_2 - a_1)(b_2 - b_1)},$$

$$B = \frac{[\varphi_3(a_2) - \varphi_3(a_1)]b_2 - [\varphi_4(a_2) - \varphi_4(a_1)]b_1}{(a_2 - a_1)(b_2 - b_1)},$$

$$C = \frac{[\psi_3(b_2) - \psi_3(b_1)]a_2 - [\psi_4(b_2) - \psi_4(b_1)]a_1}{(a_2 - a_1)(b_2 - b_1)},$$

$$D = \frac{[\varphi_4(a_2) - \varphi_4(a_1)] - [\varphi_3(a_2) - \varphi_3(a_1)]}{(a_2 - a_1)(b_2 - b_1)}.$$

Let

$$\bar{\psi}_3(y) = \psi_3(y) - u_0(a_1, y), \qquad \bar{\psi}_4(y) = \psi_4(y) - u_0(a_2, y),$$
$$\bar{\varphi}_3(x) = \varphi_3(x) - u_0(x, b_1), \qquad \bar{\varphi}_4(x) = \varphi_4(x) - u_0(x, b_2).$$

Evidently, we have

$$\bar{\psi}_j(b_1) = \bar{\psi}_j(b_2) = \bar{\varphi}_j(a_1) = \bar{\varphi}_j(a_2) = 0, \qquad j = 3, 4.$$

We can show that the problem (2.3.1)–(2.3.5) is ill-posed in the classical sense. Really, the functions

$$u_m(x, y) = \varepsilon \cdot \frac{\bar{a}_m\Big(\dfrac{3\pi}{4}\Big)\,\bar{b}_m(y) - \bar{a}_m(y)\,\bar{b}_m\Big(\dfrac{3\pi}{4}\Big)}{\bar{a}_m\Big(\dfrac{3\pi}{4}\Big)\,\bar{b}_m\Big(\dfrac{\pi}{4}\Big) - \bar{a}_m\Big(\dfrac{\pi}{4}\Big)\,\bar{b}_m\Big(\dfrac{3\pi}{4}\Big)} \cdot \cos 2(2m + 1)x, \tag{2.3.6}$$

where

$$\begin{aligned}\bar{a}_m(y) &= \sinh M\pi\left[\sinh M(\pi-y) + My\cosh M(\pi-y)\right]\\&\quad - M\pi\left[\sinh My + M(\pi-y)\cosh My\right],\\\bar{b}_m(y) &= \sinh M\pi\sinh My - M^2\pi y\cosh M(\pi-y)\\&\quad + My\left[\pi\cosh M\pi\sinh My - y\sinh M\pi\cosh My\right],\\M &= 2(2m+1),\end{aligned}$$

are the solutions of the problem (2.3.1)–(2.3.5) with $\psi_j(y) = 0$, $j = 1, 2, 3, 4$, $\varphi_1(x) = \varphi_2(x) = \varphi_4(x) = 0$, $\varphi_3(x) = \varepsilon\cos 2(2m+1)x$, $a_3 = b_3 = \pi$, $a_1 = b_1 = \pi/4$, and $a_2 = b_2 = 3\pi/4$.

From (2.3.6) it follows that for each pair of spaces of the variable x and for any constants $\bar{\varepsilon} > 0$ and $\bar{C} > 0$, we can choose ε and m such that

$$\|\varepsilon\cos 2(2m+1)x\| < \bar{\varepsilon}, \qquad \|u_m(x,y)\| > \bar{C}.$$

Hence, we see that the problem in question is ill-posed in the classical sense.

Let us consider the uniqueness of solution of the problem (2.3.1)–(2.3.5). For simplicity, we set

$$\psi_1(y) = \psi_2(y) = 0, \qquad \varphi_1(x) = \varphi_2(x) = 0.$$

A function $u(x,y)$ biharmonic in D_1 and satisfying the condition $\left.\frac{\partial u}{\partial \nu}\right|_{S_1} = 0$ can be represented in the form

$$\begin{aligned}u(x,y) &= \alpha_0 + \beta_0 B_0(y) + c_0 C_0(x)\\&\quad + \sum_{m=1}^{\infty}\left\{\left[\alpha_m A_m(y) + \beta_m B_m(y)\right]\cos\frac{m\pi x}{a_3}\right.\\&\qquad\left. + \left[c_m C_m(x) + d_m D_m(x)\right]\cos\frac{m\pi y}{b_3}\right\}, \qquad (2.3.7)\end{aligned}$$

where

$$B_0(y) = \frac{2}{b_3^3}y^3 - \frac{3}{b_3^2}y^2, \qquad C_0(x) = \frac{2}{a_3^3}x^3 - \frac{3}{a_3^2}x^2,$$

$$\begin{aligned}A_m(y) &= \left[\sinh\frac{m\pi b_3}{a_3}\sinh\frac{m\pi(b_3-y)}{a_3}\right.\\&\quad\left. + \frac{m\pi y}{a_3}\cosh\frac{m\pi(b_3-y)}{a_3} - \frac{m\pi b_3}{a_3}\left(\sinh\frac{m\pi y}{a_3} + \frac{m\pi(b_3-y)}{a_3}\cosh\frac{m\pi y}{a_3}\right)\right]\Xi,\end{aligned} \qquad (2.3.8)$$

$$B_m(y) = \left[\sinh\frac{m\pi b_3}{a_3}\sinh\frac{m\pi y}{a_3} + \frac{m\pi}{a_3}\left(b_3\cosh\frac{m\pi b_3}{a_3}\sinh\frac{m\pi y}{a_3} - y\sinh\frac{m\pi b_3}{a_3}\cosh\frac{m\pi y}{a_3}\right) - \frac{m^2\pi^2 b_3 y}{a_3^2}\cosh\frac{m\pi(b_3-y)}{a_3}\right]\Xi, \tag{2.3.9}$$

$$C_m(x) = \left[\sinh\frac{m\pi a_3}{b_3}\sinh\frac{m\pi(a_3-x)}{b_3} + \frac{m\pi x}{b_3}\cosh\frac{m\pi(a_3-x)}{b_3} - \frac{m\pi a_3}{b_3}\left(\sinh\frac{m\pi x}{b_3} + \frac{m\pi(a_3-x)}{b_3}\cosh\frac{m\pi x}{b_3}\right)\right]\Theta, \tag{2.3.10}$$

$$D_m(x) = \left[\sinh\frac{m\pi a_3}{b_3}\sinh\frac{m\pi x}{b_3} + \frac{m\pi}{b_3}\left(a_3\cosh\frac{m\pi a_3}{b_3}\sinh\frac{m\pi x}{b_3} - x\sinh\frac{m\pi a_3}{b_3}\cosh\frac{m\pi x}{b_3}\right) - \frac{m^2\pi^2 a_3 x}{b_3^2}\cosh\frac{m\pi(a_3-x)}{b_3}\right]\Theta, \tag{2.3.11}$$

$$\Xi = \left(\sinh^2\frac{m\pi b_3}{a_3} - \frac{m^2\pi^2 b_3^2}{a_3^2}\right)^{-1}, \qquad \Theta = \left(\sinh^2\frac{m\pi a_3}{b_3} - \frac{m^2\pi^2 a_3^2}{b_3^2}\right)^{-1},$$

$$\alpha_m, \beta_m, c_m, d_m \quad \text{are arbitrary constants.} \tag{2.3.12}$$

The following theorem holds.

Theorem 2.3.1. *If*

$$|\alpha_m|, |\beta_m|, |c_m|, |d_m| < C \tag{2.3.13}$$

where C is some number, then the equalities

$$u(a_1, y) = u(a_2, y) = 0, \qquad b_1 \le y \le b_2, \tag{2.3.14}$$

$$u(x, b_1) = u(x, b_2) = 0, \qquad a_1 \le x \le a_2, \tag{2.3.15}$$

imply that $u(x, y) = 0$ in D_1.

Proof. Here we use some results from the theory of Dirichlet series (Leontiev, 1976; Shishatskii, 1979).

By (2.3.14), (2.3.15), the functions

$$\psi_j(y) = \alpha_0 + \beta_0 B_0(y) + c_0 C_0(a_{j-2})$$

$$+\sum_{m=1}^{\infty}\Big\{\big[\alpha_m A_m(y)+\beta_m B_m(y)\big]\cos\frac{m\pi a_{j-2}}{a_3} + \big[c_m C_m(a_{j-2})+d_m D_m(a_{j-2})\big]\cos\frac{m\pi y}{b_3}\Big\},\qquad j=3,4,$$

vanish on the segment $[b_1, b_2]$; and the functions

$$\varphi_j(x)=\alpha_0+\beta_0 B_0(b_{j-2})+c_0C_0(x) +\sum_{m=1}^{\infty}\Big\{\big[\alpha_m A_m(b_{j-2})+\beta_m B_m(b_{j-2})\big]\cos\frac{m\pi x}{a_3} + \big[c_m C_m(x)+d_m D_m(x)\big]\cos\frac{m\pi b_{j-2}}{b_3}\Big\},\qquad j=3,4,$$

vanish on the segment $[a_1, a_2]$.

From (2.3.13) it follows that the functions $\psi_j(y)$, $\varphi_j(x)$ can be continued analytically onto the rectangles

$$\overline{P}=\Big\{z=y+\mathrm{i}\xi\ \Big|\ -\frac{a_3d}{\pi}\le \operatorname{Re}z\le\frac{a_3d}{\pi},\ -\frac{b_3d}{\pi}\le\operatorname{Im}z\le\frac{b_3d}{\pi}\Big\},$$

$$\overline{\Pi}=\Big\{\zeta=x+\mathrm{i}\eta\ \Big|\ -\frac{b_3d}{\pi}\le \operatorname{Re}\zeta\le\frac{b_3d}{\pi},\ -\frac{a_3d}{\pi}\le\operatorname{Im}\zeta\le\frac{a_3d}{\pi}\Big\},$$

$$d>2\max\Big\{\frac{a_3}{b_3},\frac{b_3}{a_3}\Big\}\pi,$$

by the equalities

$$\psi_j(z)=a^{[0]}_{j1,0}+a^{[0]}_{j1,2}z^2+a^{[0]}_{j1,3}z^3 +\sum_{m=1}^{\infty}\Big[a^{[1]}_{jm,0}\mathrm{e}^{(m\pi/a_3)z}+a^{[1]}_{jm,1}z\mathrm{e}^{(m\pi/a_3)z}+a^{[2]}_{jm,0}\mathrm{e}^{-(m\pi/a_3)z} + a^{[2]}_{jm,1}z\mathrm{e}^{-(m\pi/a_3)z}+a^{[3]}_{jm,0}\mathrm{e}^{\mathrm{i}(m\pi/b_3)z}+a^{[4]}_{jm,0}\mathrm{e}^{-\mathrm{i}(m\pi/b_3)z}\Big], \tag{2.3.16}$$

$$\varphi_j(\zeta)=b^{[0]}_{j1,0}+b^{[0]}_{j1,2}\zeta^2+b^{[0]}_{j1,3}\zeta^3 +\sum_{m=1}^{\infty}\Big[b^{[1]}_{jm,0}\mathrm{e}^{(m\pi/b_3)\zeta}+b^{[1]}_{jm,1}\zeta\mathrm{e}^{(m\pi/b_3)\zeta}+b^{[2]}_{jm,0}\mathrm{e}^{-(m\pi/b_3)\zeta} + b^{[2]}_{jm,1}\zeta\mathrm{e}^{-(m\pi/b_3)\zeta}+b^{[3]}_{jm,0}\mathrm{e}^{\mathrm{i}(m\pi/a_3)\zeta}+b^{[4]}_{jm,0}\mathrm{e}^{-\mathrm{i}(m\pi/a_3)\zeta}\Big]. \tag{2.3.17}$$

Here

$$a^{[0]}_{j1,0} = \alpha_0 + c_0 C_0(a_{j-2}), \qquad a^{[0]}_{j1,2} = \frac{3\beta_0}{b_3^2}, \qquad a^{[0]}_{j1,3} = \frac{3\beta_0}{b_3^3},$$

$$a^{[1]}_{jm,0} = \bar{K}_{1m}\Big[-\alpha_m\Big(\sinh\frac{m\pi b_3}{a_3}\cdot e^{-(m\pi/a_3)b_3} + \frac{m\pi b_3}{a_3}\Big(1+\frac{m\pi b_3}{a_3}\Big)\Big) - \beta_m\Big(\sinh\frac{m\pi b_3}{a_3} + \frac{m\pi b_3}{a_3}\cosh\frac{m\pi b_3}{a_3}\Big)\Big]\cos\frac{m\pi a_{j-2}}{a_3},$$

$$a^{[1]}_{jm,1} = \bar{K}_{1m}\Big[-\alpha_m\Big(\sinh\frac{m\pi b_3}{a_3}\cdot e^{-(m\pi/a_3)b_3} + \frac{m\pi b_3}{a_3}\Big)\frac{m\pi}{a_3} - \beta_m\Big(\sinh\frac{m\pi b_3}{a_3} + \frac{m\pi b_3}{a_3}e^{(m\pi/a_3)b_3} + \frac{m\pi}{a_3}\Big)\Big]\cos\frac{m\pi a_{j-2}}{a_3},$$

$$a^{[2]}_{jm,0} = \bar{K}_{1m}\Big[-\alpha_m\Big(\sinh\frac{m\pi b_3}{a_3}\cdot e^{(m\pi/a_3)b_3} + \frac{m\pi b_3}{a_3}\Big(1-\frac{m\pi b_3}{a_3}\Big)\Big) - \beta_m\Big(\sinh\frac{m\pi b_3}{a_3} + \frac{m\pi b_3}{a_3}\cosh\frac{m\pi b_3}{a_3}\Big)\Big]\cos\frac{m\pi a_{j-2}}{a_3},$$

$$a^{[2]}_{jm,1} = \bar{K}_{1m}\Big[\alpha_m\Big(\sinh\frac{m\pi b_3}{a_3}\cdot e^{(m\pi/a_3)b_3} + \frac{m\pi b_3}{a_3}\Big)\frac{m\pi}{a_3} - \beta_m\Big(\sinh\frac{m\pi b_3}{a_3} + \frac{m\pi b_3}{a_3}e^{(m\pi/a_3)b_3} + \frac{m\pi}{a_3}\Big)\Big]\cos\frac{m\pi a_{j-2}}{a_3},$$

$$a^{[3]}_{jm,0} = a^{[4]}_{jm,0} = \frac{c_m C_m(a_{j-2}) + d_m D_m(a_{j-2})}{2},$$

$$b^{[0]}_{j1,0} = \alpha_0 + \beta_0 B_0(b_{j-2}), \qquad b^{[0]}_{j1,2} = \frac{-2c_0}{a_3^2}, \qquad b^{[0]}_{j1,3} = \frac{3c_0}{a_3^3},$$

$$b^{[1]}_{jm,0} = \bar{K}_{2m}\Big[-c_m\Big(\sinh\frac{m\pi a_3}{b_3}\cdot e^{-(m\pi/b_3)a_3} + \frac{m\pi a_3}{b_3}\Big(1+\frac{m\pi a_3}{b_3}\Big)\Big) + d_m\Big(\sinh\frac{m\pi a_3}{b_3} + \frac{m\pi a_3}{b_3}\cosh\frac{m\pi a_3}{b_3}\Big)\Big]\cos\frac{m\pi b_{j-2}}{b_3},$$

$$b^{[1]}_{jm,1} = \bar{K}_{2m}\Big[c_m\Big(\sinh\frac{m\pi a_3}{b_3}\cdot e^{-(m\pi/b_3)a_3} + \frac{m\pi a_3}{b_3}\Big)\frac{m\pi}{b_3} - d_m\Big(\sinh\frac{m\pi a_3}{b_3} + \frac{m\pi a_3}{b_3}e^{(m\pi/b_3)a_3}\Big)\frac{m\pi}{b_3}\Big]\cos\frac{m\pi b_{j-2}}{b_3},$$

$$b^{[2]}_{jm,0} = \bar{K}_{2m}\Big[c_m\Big(\sinh\frac{m\pi a_3}{b_3}\cdot e^{(m\pi/b_3)a_3} - \frac{m\pi a_3}{b_3}\Big(1-\frac{m\pi a_3}{b_3}\Big)\Big)$$
$$- d_m\Big(\sinh\frac{m\pi a_3}{b_3} + \frac{m\pi a_3}{b_3}\cosh\frac{m\pi a_3}{b_3}\Big)\Big]\cos\frac{m\pi b_{j-2}}{b_3},$$

$$b^{[2]}_{jm,1} = \bar{K}_{2m}\Big[c_m\Big(\sinh\frac{m\pi a_3}{b_3}\cdot e^{(m\pi/b_3)a_3} + \frac{m\pi a_3}{b_3}\Big)\frac{m\pi}{b_3}$$
$$- d_m\Big(\sinh\frac{m\pi a_3}{b_3} + \frac{m\pi a_3}{b_3}\cosh\frac{m\pi a_3}{b_3}\Big)\frac{m\pi}{b_3}\Big]\cos\frac{m\pi b_{j-2}}{b_3},$$

$$b^{[3]}_{jm,0} = b^{[4]}_{jm,0} = \frac{\alpha_m A_m(b_{j-2}) + \beta_m B_m(b_{j-2})}{2}, \qquad j = 3,4,$$

$$\bar{K}_{1m} = \Big(\sinh^2\frac{m\pi b_3}{a_3} - \frac{m^2\pi^2 b_3^2}{a_3^2}\Big)^{-1}, \qquad \bar{K}_{2m} = \Big(\sinh^2\frac{m\pi a_3}{b_3} - \frac{m^2\pi^2 a_3^2}{b_3^2}\Big)^{-1}.$$

Since $\psi_j(z) = 0$, $j = 3,4$, on the real segment $[b_1, b_2]$ and $\varphi_j(\zeta) = 0$, $j = 3,4$, on the real segment $[a_1, a_2]$, therefore, by uniqueness of analytical continuation of a function of one complex variable, we have $\psi_j(z) = 0$, $j = 3,4$, in $\overline{P}$ and $\varphi_j(\zeta) = 0$, $j = 3,4$, in $\overline{\Pi}$.

We introduce the function

$$Q_1(\lambda) = \lambda\sinh\lambda b_3\sin^2\lambda a_3.$$

The function $Q_1(\lambda)$ is an entire function of exponential type (see Leontiev, 1976), which at the points

$$\lambda^{[0]}_1 = 0, \qquad \lambda^{[1]}_m = \frac{m\pi}{a_3}, \qquad \lambda^{[2]}_m = -\frac{m\pi}{a_3},$$
$$\lambda^{[3]}_m = i\,\frac{m\pi}{b_3}, \qquad \lambda^{[4]}_m = -i\,\frac{m\pi}{b_3}, \qquad m = 1,2,\ldots,$$

has zeros of the orders

$$\nu^{[0]}_1 = 4, \qquad \nu^{[1]}_m = 2, \qquad \nu^{[2]}_m = 2, \qquad \nu^{[3]}_m = 1, \qquad \nu^{[4]}_m = 1,$$
$$m = 1,2,\ldots.$$

By immediate computation we obtain

$$Q_1(\lambda) = \frac{1}{2}\sum_{n=0}^{\infty}\big[b_3^{2n+1} - r_1^{2n+1}\cos(2n+1)\psi_0\big]\frac{\lambda^{2n+2}}{(2n+1)!},$$

where $r_1 = \sqrt{b_3^2 + 4a_3^2}$, $\psi_0 = \arctan(2a_3/b_3)$.

Let us denote the Borel functional transform of $Q_1(\lambda)$ (see Leontiev, 1976) by $\gamma_1(z)$. Then we have

$$\gamma_1(z) = \frac{1}{2}\sum_{n=0}^{\infty}\left[b_3^{2n+1} - r_1^{2n+1}\cos(2n+1)\psi_0\right]\frac{2n+2}{z^{2n+3}}$$
$$= \frac{b_3 z}{(z^2-b_3^2)^2} - \frac{r_1 z(z^4 - 2r_1^2 z^2 + (2\cos 2\psi_0 - 1)r_1^4)^2}{(z^4 - 2r_1^2 z^2\cos 2\psi_0 + r_1^4)^2}\cos\psi_0.$$

The function $\gamma_1(z)$ has singularities only at the points

$$z = b_3, \quad -b_3, \quad b_3 + 2a_3\mathrm{i}, \quad b_3 - 2a_3\mathrm{i}, \quad -b_3 + 2a_3\mathrm{i}, \quad -b_3 - 2a_3\mathrm{i}.$$

The smallest convex domain which contains all these singularities is the rectangle

$$\overline{P_1} = \left\{z \mid -b_3 \le \operatorname{Re} z \le b_3, \ -2a_3 \le \operatorname{Im} z \le 2a_3\right\}.$$

By condition (2.3.13) of Theorem 2.3.1, we have $P \supset \overline{P_1}$.

We will denote by $A_{m,k}^{[i]}(t)$ the Borel functional transform of the entire function $Q_1(\lambda)/(\lambda - \lambda_m^{[i]})^{k+1}$; $k = 0, 1, \dots, \nu_m^{[i]} - 1$; $i = 0, 1, 2, 3, 4$. We set (Leontiev, 1976)

$$M_{1m}^{[i]}(\mu) = \frac{(\mu - \lambda_m^{[i]})^{\nu_m^{[i]}}}{Q_1(\mu)},$$

$$\psi_{m,k}^{[i]}(t) = \frac{1}{k!\,(\nu_m^{[i]} - k - 1)!}\sum_{q=0}^{\nu_m^{[i]}-k-1} q!\cdot C_{\nu_m^{[i]}-k-1}^{q}\left[M_{1m}^{[i]}(\mu)\right]_{\mu=\lambda_m^{[i]}}^{(\nu_m^{[i]}-k-q-1)} A_{m,q}^{[i]}(t).$$

Now we show that

$$\frac{1}{2\pi\mathrm{i}}\int_{C_1} t^s \mathrm{e}^{\lambda_p^{[i]}t}\psi_{m,k}^{[n]}(t)\,\mathrm{d}t = \begin{cases} 1, & i = n,\ p = m,\ s = k, \\ 0, & \text{otherwise}, \end{cases} \tag{2.3.18}$$

where C_1 is a simple closed contour lying in $P \setminus \overline{P_1}$. Really, we have (we will denote by J the left-hand side of equality (2.3.18))

$$J = \frac{1}{k!\,(\nu_m^{[n]} - k - 1)!}\sum_{q=0}^{\nu_m^{[n]}-k-1} q!\cdot C_{\nu_m^{[n]}-k-1}^{q}\left[M_{1m}^{[n]}(\mu)\right]_{\mu=\lambda_m^{[n]}}^{(\nu_m^{[n]}-k-q-1)}$$
$$\times \frac{1}{2\pi\mathrm{i}}\int_{C_1} t^s \mathrm{e}^{\lambda_p^{[i]}t} A_{m,q}^{[n]}(t)\,\mathrm{d}t.$$

Since

$$\frac{1}{2\pi \mathrm{i}} \int_{C_1} \mathrm{e}^{\lambda t} A^{[n]}_{m,q}(t)\,\mathrm{d}t = \frac{Q_1(\lambda)}{(\lambda - \lambda^{[n]}_m)^{q+1}},$$

therefore,

$$\frac{1}{2\pi \mathrm{i}} \int_{C_1} t^s \mathrm{e}^{\lambda^{[i]}_p t} A^{[n]}_{m,q}(t)\,\mathrm{d}t = \left[\frac{Q_1(\lambda)}{(\lambda - \lambda^{[n]}_m)^{q+1}}\right]^{(s)}_{\lambda=\lambda^{[i]}_p}.$$

However,

$$q!\,\frac{Q_1(\lambda)}{(\lambda - \lambda^{[n]}_m)^{q+1}} = \left[\frac{Q_1(\lambda) - Q_1(\mu)}{\lambda - \mu}\right]^{(q)}_{\mu=\lambda^{[n]}_m}.$$

Therefore,

$$q!\,\frac{1}{2\pi \mathrm{i}} \int_{C_1} t^s \mathrm{e}^{\lambda^{[i]}_p t} A^{[n]}_{m,q}(t)\,\mathrm{d}t = \left\{\left[\frac{Q_1(\lambda) - Q_1(\mu)}{\lambda - \mu}\right]^{(q)}_{\mu=\lambda^{[n]}_m}\right\}^{(s)}_{\lambda=\lambda^{[i]}_p}$$

$$= \left\{\left[\frac{Q_1(\lambda) - Q_1(\mu)}{\lambda - \mu}\right]^{(s)}_{\lambda=\lambda^{[i]}_p}\right\}^{(q)}_{\mu=\lambda^{[n]}_m} = \left[s!\,\frac{Q_1(\mu)}{(\mu - \lambda^{[i]}_p)^{s+1}}\right]^{(q)}_{\mu=\lambda^{[n]}_m}.$$

Hence, we see that

$$\begin{aligned} J &= \frac{s!}{k!\,(\nu^{[n]}_m - k - 1)!} \\ &\quad\times \sum_{q=0}^{\nu^{[n]}_m - k - 1} C^q_{\nu^{[n]}_m - k - 1} [M^{[n]}_{1m}(\mu)]^{(\nu^{[n]}_m - k - q - 1)}_{\mu=\lambda^{[n]}_m} \left[\frac{Q_1(\mu)}{(\mu - \lambda^{[i]}_p)^{s+1}}\right]^{(q)}_{\mu=\lambda^{[n]}_m} \\ &= \frac{s!}{k!\,(\nu^{[n]}_m - k - 1)!} \left[M^{[n]}_{1m}(\mu)\,\frac{Q_1(\mu)}{(\mu - \lambda^{[i]}_p)^{s+1}}\right]^{(\nu^{[n]}_m - k - 1)}_{\mu=\lambda^{[n]}_m} \\ &= \frac{s!}{k!\,(\nu^{[n]}_m - k - 1)!} \left[\frac{(\mu - \lambda^{[n]}_m)^{\nu^{[n]}_m}}{(\mu - \lambda^{[i]}_p)^{s+1}}\right]^{(\nu^{[n]}_m - k - 1)}_{\mu=\lambda^{[n]}_m} \\ &= \begin{cases} 1, & i = n,\ p = m,\ s = k, \\ 0, & \text{otherwise,} \end{cases} \end{aligned}$$

which was to be proved.

Multiply both sides of equality (2.3.16) by the functions

$$\frac{1}{2\pi \mathrm{i}}\,\psi^{[0]}_{1,0}(z), \qquad \frac{1}{2\pi \mathrm{i}}\,\psi^{[3]}_{m,0}(z), \qquad m = 1, 2, \dots,$$

and integrate around the contour C_1. Then, taking into account (2.3.18), we obtain

$$a_{j1,0}^{[0]} = \frac{1}{2\pi \mathrm{i}} \int_{C_1} \psi_j(z)\psi_{1,0}^{[0]}(z)\,\mathrm{d}z, \qquad a_{jm,0}^{[3]} = \frac{1}{2\pi \mathrm{i}} \int_{C_1} \psi_j(z)\psi_{m,0}^{[3]}(z)\,\mathrm{d}z,$$
$$j = 3, 4, \qquad m = 1, 2, \ldots. \tag{2.3.19}$$

Now we consider the function

$$Q_2(\lambda) = \lambda \sinh \lambda a_3 \sin^2 \lambda b_3 = \frac{1}{2}\sum_{n=0}^{\infty}\left[a_3^{2n+1} - r_2^{2n+1}\cos(2n+1)\varphi_0\right]\frac{\lambda^{2n+2}}{(2n+1)!}$$

where $r_2 = \sqrt{a_3^2 + 4b_3^2}$, $\varphi_0 = \arctan(2b_3/a_3)$. The function $Q_2(\lambda)$ is an entire function of exponential type; at the points

$$\tilde{\lambda}_1^{[0]} = 0, \qquad \tilde{\lambda}_m^{[1]} = \frac{m\pi}{b_3}, \qquad \tilde{\lambda}_m^{[2]} = -\frac{m\pi}{b_3},$$
$$\tilde{\lambda}_m^{[3]} = \mathrm{i}\,\frac{m\pi}{a_3}, \qquad \tilde{\lambda}_m^{[4]} = -\mathrm{i}\,\frac{m\pi}{a_3}, \qquad m = 1, 2, \ldots,$$

it has zeros of the orders

$$\tilde{\nu}_1^{[0]} = 4, \qquad \tilde{\nu}_m^{[1]} = \tilde{\nu}_m^{[2]} = 2, \qquad \tilde{\nu}_m^{[3]} = \tilde{\nu}_m^{[4]} = 1, \qquad m = 1, 2, \ldots,$$

respectively.

We will denote by $\gamma_2(\zeta)$ the Borel functional transform of $Q_2(\lambda)$. Then

$$\begin{aligned}\gamma_2(\zeta) &= \frac{1}{2}\sum_{n=0}^{\infty}\left[a_3^{2n+1} - r_2^{2n+1}\cos(2n+1)\varphi_0\right]\frac{2n+2}{\zeta^{2n+3}} \\ &= \frac{a_3\zeta}{(\zeta^2 - a_3^2)^2} - \frac{r_2\zeta(\zeta^4 - 2r_2^2\zeta^2 + (2\cos 2\varphi_0 - 1)r_2^4)}{(\zeta^4 - 2r_2^2\zeta^2\cos 2\varphi_0 + r_2^4)^2}\cos\varphi_0,\end{aligned}$$

where $r_2 = \sqrt{a_3^2 + 4b_3^2}$, $\varphi_0 = \arctan(2b_3/a_3)$.

This function has singularities only at the points

$$\zeta = a_3, \quad -a_3, \quad a_3 + 2b_3\mathrm{i}, \quad a_3 - 2b_3\mathrm{i}, \quad -a_3 + 2b_3\mathrm{i}, \quad -a_3 - 2b_3\mathrm{i}.$$

The smallest convex domain which contains all these singularities of $\gamma_2(\zeta)$ is the rectangle

$$\overline{\Pi_1} = \{\zeta \mid -a_3 \le \operatorname{Re}\zeta \le a_3,\ -2b_3 \le \operatorname{Im}\zeta \le 2b_3\}.$$

From condition (2.3.13) of Theorem 2.3.1 it follows that $\overline{\Pi_1} \subset \Pi$.

We suppose now that C_2 is a simple closed contour lying in $\Pi \setminus \overline{\Pi}_1$.

Let $B^{[i]}_{m,k}(t)$ be the Borel functional transform of the entire function $Q_2(\lambda)/(\lambda - \tilde{\lambda}^{[i]}_m)^{k+1}$; $k = 0, 1, \dots, \tilde{\nu}^{[i]}_m - 1$; $i = 0, 1, 2, 3, 4$. Further, we set

$$M^{[i]}_{2m}(\mu) = \frac{(\mu - \tilde{\lambda}^{[i]}_m)^{\tilde{\nu}^{[i]}_m}}{Q_2(\mu)},$$

$$\varphi^{[i]}_{m,k}(t) = \frac{1}{k!\,(\tilde{\nu}^{[i]}_m - k - 1)!} \sum_{q=0}^{\tilde{\nu}^{[i]}_m - k - 1} q! \cdot C^q_{\tilde{\nu}^{[i]}_m - k - 1} [M^{[i]}_{2m}(\mu)]^{(\tilde{\nu}^{[i]}_m - k - q - 1)}_{\mu = \tilde{\lambda}^{[i]}_m} B^{[i]}_{m,q}(t).$$

Analogously to (2.3.18), we show that

$$\frac{1}{2\pi \mathrm{i}} \int_{C_2} t^s \mathrm{e}^{\tilde{\lambda}^{[i]}_p t} \varphi^{[n]}_{m,k}(t)\,\mathrm{d}t = \begin{cases} 1, & i = n,\ p = m,\ s = k, \\ 0, & \text{otherwise.} \end{cases} \tag{2.3.20}$$

Multiply both sides of (2.3.17) by the functions

$$\frac{1}{2\pi \mathrm{i}}\, \varphi^{[0]}_{1,0}(\zeta), \qquad \frac{1}{2\pi \mathrm{i}}\, \varphi^{[3]}_{m,0}(\zeta), \qquad m = 1, 2, \dots,$$

and integrate around the contour C_2. Then, taking into account (2.3.20), we obtain

$$b^{[0]}_{j1,0} = \frac{1}{2\pi \mathrm{i}} \int_{C_2} \varphi_j(\zeta) \varphi^{[0]}_{1,0}(\zeta)\,\mathrm{d}\zeta, \qquad b^{[3]}_{jm,0} = \frac{1}{2\pi \mathrm{i}} \int_{C_2} \varphi_j(\zeta) \varphi^{[3]}_{m,0}(\zeta)\,\mathrm{d}\zeta,$$

$$j = 3, 4, \qquad m = 1, 2, \dots. \tag{2.3.21}$$

Taking into account that $\psi_j(z) = 0$, $j = 3, 4$, in $\overline{P}$ and $\varphi_j(\zeta) = 0$, $j = 3, 4$, in $\overline{\Pi}$, we obtain from (2.3.19), (2.3.20)

$$a^{[0]}_{j1,0} = b^{[0]}_{j1,0} = a^{[3]}_{jm,0} = b^{[3]}_{jm,0} = 0, \qquad j = 3, 4, \quad m = 1, 2, \dots.$$

Hence it follows that $\alpha_m = \beta_m = c_m = d_m = 0$ for all $m = 0, 1, 2, \dots$.

Returning to formula (2.3.7), we see that $u(x, y) = 0$ in D_1. Thus, Theorem 2.3.1 is proved. □

Suppose now that

$$a_1 = \frac{a(2k+1)}{2}, \qquad a_2 = \frac{a(2m+1)}{2}, \qquad a_3 = sa, \qquad 0 < a < \infty,$$

$$b_1 = \frac{b(2r+1)}{2}, \qquad b_2 = \frac{b(2p+1)}{2}, \qquad b_3 = qb, \qquad 0 < b < \infty,$$

$$0 \le k < m < s < \infty, \qquad 0 \le r < p < q < \infty,$$

where k, m, s, r, p, and q are integers.

We assume that

$$\psi_j(y) \not\equiv 0, \qquad \varphi_j(x) \not\equiv 0, \qquad j = 1, 2, 3, 4,$$
$$\int_0^{a_3} \varphi_j(x)\,\mathrm{d}x = \int_{a_1}^{a_2} \varphi_{j+2}(x)\,\mathrm{d}x = 0,$$
$$\int_0^{b_3} \psi_j(y)\,\mathrm{d}y = \int_{b_1}^{b_2} \psi_{j+2}(y)\,\mathrm{d}y = 0, \qquad j = 1, 2.$$

Then the solution of the problem (2.3.1)–(2.3.5) can be represented as follows

$$u(x, y) = u_1(x, y) + u_2(x, y),$$

where $u_1(x, y)$ and $u_2(x, y)$ are the solutions of the problems

$$\Delta^2 u_1(x, y) = 0 \quad \text{in} \quad D_1, \tag{2.3.22}$$
$$u'_{1x}(0, y) = 0, \qquad u'_{1x}(a_3, y) = 0, \qquad 0 \le y \le b_3, \tag{2.3.23}$$
$$u'_{1y}(x, 0) = \varphi_1(x), \qquad u'_{1y}(x, b_3) = \varphi_2(x), \qquad 0 \le x \le a_3, \tag{2.3.24}$$
$$u_1(a_1, y) = 0, \qquad u_1(a_2, y) = 0, \qquad b_1 \le y \le b_2, \tag{2.3.25}$$
$$u_1(x, b_1) = \varphi_3(x), \qquad u_1(x, b_2) = \varphi_4(x), \qquad a_1 \le x \le a_2, \tag{2.3.26}$$
$$\Delta^2 u_2(x, y) = 0 \quad \text{in} \quad D_1, \tag{2.3.27}$$
$$u'_{2x}(0, y) = \psi_1(y), \qquad u'_{2x}(a_3, y) = \psi_2(y), \qquad 0 \le y \le b_3, \tag{2.3.28}$$
$$u'_{2y}(x, 0) = 0, \qquad u'_{2y}(x, b_3) = 0, \qquad 0 \le x \le a_3, \tag{2.3.29}$$
$$u_2(a_1, y) = \psi_3(y), \qquad u_2(a_2, y) = \psi_4(y), \qquad b_1 \le y \le b_2, \tag{2.3.30}$$
$$u_2(x, b_1) = 0, \qquad u_2(x, b_2) = 0, \qquad a_1 \le x \le a_2. \tag{2.3.31}$$

The solution of the problem (2.3.22)–(2.3.26) can be represented as

$$u_1(x, y) = \sum_{k=1}^{4} u_{1k}(x, y),$$

where $u_{1k}(x, y)$, $k = 1, 2, 3, 4$, solve the problems

$$\Delta^2 u_{1k}(x, y) = 0 \quad \text{in} \quad D_1, \tag{2.3.32}$$
$$u'_{1kx}(0, y) = 0, \qquad u'_{1kx}(a_3, y) = 0, \qquad 0 \le y \le b_3, \tag{2.3.33}$$
$$u'_{1ky}(x, 0) = \varphi_1(x), \qquad u'_{1ky}(x, b_3) = \varphi_2(x), \qquad 0 \le x \le a_3, \tag{2.3.34}$$
$$u_{1k}(a_1, y) = 0, \qquad u_{1k}(a_2, y) = 0, \qquad b_1 \le y \le b_2, \tag{2.3.35}$$

$$u_{1k}(x,b_1)=\varphi_3(x),\qquad u_{1k}(x,b_2)=\varphi_4(x),\qquad a_1\le x\le a_2,\tag{2.3.36}$$

where $\varphi_j(x)=0$ for $j\ne k$; $j,k=1,2,3,4$.

Let us consider the problem

$$\Delta^2 u_{13}(x,y)=0\quad \text{in}\quad D_1,\tag{2.3.37}$$

$$u'_{13x}(0,y)=0,\qquad u'_{13x}(a_3,y)=0,\qquad 0\le y\le b_3,\tag{2.3.38}$$

$$u'_{13y}(x,0)=0,\qquad u'_{13y}(x,b_3)=0,\qquad 0\le x\le a_3,\tag{2.3.39}$$

$$u_{13}(a_1,y)=0,\qquad u_{13}(a_2,y)=0,\qquad b_1\le y\le b_2,\tag{2.3.40}$$

$$u_{13}(x,b_1)=\varphi_3(x),\qquad u_{13}(x,b_2)=0,\qquad a_1\le x\le a_2,\tag{2.3.41}$$

and the question whether it is conditionally well-posed.

The solution of the problem (2.3.37)–(2.3.41) can be represented as

$$u_{13}(x,y)=\sum_{m=0}^{\infty} C_m A_m^{[3]}(y)\cos\lambda_m x,\tag{2.3.42}$$

where $\lambda_m=(2m+1)\pi/a$, C_m are the Fourier coefficients of the function $\varphi_3(x)$, and

$$\begin{aligned}
&A_m^{[3]}(y)\\
&\quad=\Big\{\big[\lambda_m b_2 b_3\cosh\lambda_m(b_2-b_3)-(\sinh\lambda_m b_2-\lambda_m b_2\cosh\lambda_m b_2)\sinh\lambda_m b_3\\
&\qquad\quad-\lambda_m b_3\sinh\lambda_m b_2\cosh\lambda_m b_3\big]\cosh\lambda_m y\\
&\qquad+\big[\lambda_m b_3\cosh\lambda_m b_2\cosh\lambda_m b_3+(\cosh\lambda_m b_2-b_2\sinh\lambda_m b_2)\sinh\lambda_m b_3\big]\\
&\qquad\quad\times(\sinh\lambda_m y-\lambda_m y\cosh\lambda_m y)\\
&\qquad+\lambda_m y\big[\sinh\lambda_m b_2+\lambda_m(b_3-b_2)\cosh\lambda_m b_2\big]\sinh\lambda_m b_3\sinh\lambda_m y\Big\}\\
&\quad\times\Big\{\lambda_m^2(b_2-b_1)b_3\cosh\lambda_m b_1\cosh\lambda_m b_2\cosh\lambda_m b_3\\
&\qquad+\big[b_1 b_2\sinh\lambda_m(b_2-b_1)-(b_2-b_1)b_3\sinh\lambda_m(b_2-b_1)\big]\sinh\lambda_m b_3\\
&\qquad-\lambda_m b_3\sinh\lambda_m(b_2-b_1)\cosh\lambda_m b_3\\
&\qquad-\big[\sinh\lambda_m(b_2-b_1)-\lambda_m(b_1+b_2)\cosh\lambda_m(b_2-b_1)\big]\sinh\lambda_m b_3\Big\}^{-1}.
\end{aligned}\tag{2.3.43}$$

From (2.3.42) we obtain

$$\Delta u_{13}(x,y)=\sum_{m=0}^{\infty} C_m B_m^{[3]}(y)\cos\lambda_m x,$$

where

$$
\begin{aligned}
&B_m^{[3]}(y)\\
&= 2\lambda_m^2\Big\{\big[\sinh\lambda_m b_2 - \lambda_m(b_3-b_2)\cosh\lambda_m b_3\big]\sinh\lambda_m b_3\,\cosh\lambda_m y\\
&\qquad - \big[\lambda_m b_3\cosh\lambda_m b_2\,\cosh\lambda_m b_3\\
&\qquad\qquad + (\cosh\lambda_m b_2 - b_2\sinh\lambda_m b_2)\sinh\lambda_m b_3\big]\sinh\lambda_m y\Big\}\\
&\quad\times\Big\{\lambda_m^2(b_2-b_1)b_3\cosh\lambda_m b_1\,\cosh\lambda_m b_2\,\cosh\lambda_m b_3\\
&\qquad + \big[b_1 b_2\sinh\lambda_m(b_2-b_1) - (b_2-b_1)b_3\sinh\lambda_m(b_2-b_1)\big]\sinh\lambda_m b_3\\
&\qquad - \lambda_m b_3\sinh\lambda_m(b_2-b_1)\cosh\lambda_m b_3\\
&\qquad - \big[\sinh\lambda_m(b_2-b_1) - \lambda_m(b_1+b_2)\cosh\lambda_m(b_2-b_1)\big]\sinh\lambda_m b_3\Big\}^{-1}.
\end{aligned}
\tag{2.3.44}
$$

The following theorems characterize the stability of solution of the problem (2.3.37)–(2.3.41).

Theorem 2.3.2. *If the function $u_{13}(x,y)$ is such that*

$$\|u_{13}(x,b_3)\|_{L_2(0,a_3)} \le \frac{M_1}{4}, \tag{2.3.45}$$

$$\|u_{13}(x,b_1)\|_{L_2(0,a_3)} \le \varepsilon_3, \qquad \int_{a_1}^{a_2} u_{13}(x,b_1)\,dx = 0, \tag{2.3.46}$$

$$u'_{13x}(0,y) = u'_{13x}(a_3,y) = 0, \qquad 0 \le y \le b_3, \tag{2.3.47}$$

$$u'_{13y}(x,0) = u'_{13y}(x,b_3) = 0, \qquad 0 \le x \le a_3, \tag{2.3.48}$$

$$u_{13}(a_1,y) = u_{13}(a_2,y) = 0, \qquad b_1 \le y \le b_2, \tag{2.3.49}$$

$$u_{13}(x,b_2) = 0, \qquad a_1 \le x \le a_2, \tag{2.3.50}$$

then the inequality

$$\|u_{13}(x,y)\|_{L_2(0,a_3)} \le \frac{M_1}{4}\left|\frac{A^{[3]}_{\mu(\varepsilon_3)}(y)}{A^{[3]}_{\mu(\varepsilon_3)}(b_3)}\right|$$

holds. Here $A^{[3]}_\mu(y)$ is defined by (2.3.43), and $\mu(\varepsilon_3)$ is the positive root of the equation

$$4|A^{[3]}_\mu(b_3)| = M_1\varepsilon_3^{-1}. \tag{2.3.51}$$

Note that (2.3.43) and (2.3.51) imply that $\mu(\varepsilon_3)\to\infty$ for $\varepsilon_3\to 0$.

Theorem 2.3.3. *If the function $u_{13}(x, y)$ satisfies conditions (2.3.46)–(2.3.50) and*

$$\|\Delta u_{13}(x, b_3)\|_{L_2(0,a_3)} \leq \frac{M_1}{4}, \tag{2.3.52}$$

then the inequality

$$\|u_{13}(x, y)\|_{L_2(0,a_3)} \leq \frac{M_1}{4}\left|\frac{A^{[3]}_{\bar{\mu}(\varepsilon_3)}(y)}{B^{[3]}_{\bar{\mu}(\varepsilon_3)}(b_3)}\right|$$

holds. Here $A^{[3]}_{\bar{\mu}}(y)$ and $B^{[3]}_{\bar{\mu}}(y)$ are defined by formulas (2.3.43) and (2.3.44), and $\bar{\mu}(\varepsilon_3)$ is the positive root of the equation

$$4|B^{[3]}_{\bar{\mu}}(b_3)| = M_1\varepsilon_3^{-1}. \tag{2.3.53}$$

Note that (2.3.44) and (2.3.53) imply that $\bar{\mu}(\varepsilon_3) \to \infty$ for $\varepsilon_3 \to 0$.

The above theorems are established similarly to Theorem 2.3.1. We suppose now that the function $\varphi_3(x)$ is known with accuracy δ, i. e., we know $\varphi_{3\delta}(x)$ such that

$$\|\varphi_3(x) - \varphi_{3\delta}(x)\|_{L_2(a_1,a_2)} \leq \delta.$$

We assume that the problem (2.3.37)–(2.3.41) is set conditionally well-posed, and the well-posedness set is defined by inequality (2.3.45). As an approximate solution to the problem (2.3.37)–(2.3.41) we take the function

$$u_{13N\delta}(x, y) = \sum_{m=0}^{N} \bar{C}_m A^{[3]}_m(y) \cos \lambda_m x,$$

where $\bar{C}_m$ are the Fourier coefficients of the function $\varphi_{3\delta}(x)$.

Now estimate $u_{13}(x, y) - u_{13N\delta}(x, y)$ in the metric of the space $L_2(0, a_3)$. Analogously to (2.1.37), we have

$$\|u_{13}(x, y) - u_{13N\delta}(x, y)\|_{L_2(0,a_3)} \leq |A^{[3]}_N(y)|\delta + \left|\frac{A^{[3]}_{N+1}(y)}{A^{[3]}_{N+1}(b_3)}\right|.$$

Now we suppose that the well-posedness set is defined by inequality (2.3.52). In this case we obtain

$$\|u_{13}(x, y) - u_{13N\delta}(x, y)\|_{L_2(0,a_3)} \leq |A^{[3]}_N(y)|\delta + \left|\frac{A^{[3]}_{N+1}(y)}{B^{[3]}_{N+1}(b_3)}\right|.$$

Thus, the problem (2.3.37)–(2.3.41) has been studied in the space $L_2(0,a_3)$. Analogously we can study the problems (2.3.32)–(2.3.36).

Now, for the problem (2.3.22)–(2.3.26), we formulate the following results.

Theorem 2.3.4. *If the function $u_1(x,y)$ is such that*

$$\|u_1(x,b_3)\|_{L_2(0,a_3)} \le M_1, \tag{2.3.54}$$

$$\|u'_{1y}(x,0)\|_{L_2(0,a_3)} \le \varepsilon_1, \qquad \|u'_{1y}(x,b_3)\|_{L_2(0,a_3)} \le \varepsilon_2, \tag{2.3.55}$$

$$\|u_1(x,b_1)\|_{L_2(a_1,a_2)} \le \varepsilon_3, \qquad \|u_1(x,b_2)\|_{L_2(a_1,a_2)} \le \varepsilon_4, \tag{2.3.56}$$

$$\int_0^{a_3} u'_{1y}(x,0)\,\mathrm{d}x = \int_0^{a_3} u'_{1y}(x,b_3)\,\mathrm{d}x = 0, \tag{2.3.57}$$

$$\int_{a_1}^{a_2} u_1(x,b_1)\,\mathrm{d}x = \int_{a_1}^{a_2} u_1(x,b_2)\,\mathrm{d}x = 0, \tag{2.3.58}$$

$$u'_{1x}(0,y) = u'_{1x}(a_3,y) = 0, \qquad 0 \le y \le b_3, \tag{2.3.59}$$

$$u_1(a_1,y) = u_1(a_2,y) = 0, \qquad b_1 \le y \le b_2, \tag{2.3.60}$$

then the inequality

$$\|u_1(x,y)\|_{L_2(0,a_3)} \le \frac{M_1}{4}\sum_{j=1}^{4}\left|\frac{A^{[j]}_{\mu(\varepsilon_j)}(y)}{A^{[j]}_{\mu(\varepsilon_j)}(b_3)}\right|$$

holds true. Here $A^{[j]}_{\mu}(y)$ are known functions and

$$\left|\frac{A^{[j]}_{\mu}(y)}{A^{[j]}_{\mu}(b_3)}\right| \to 0$$

as $\mu \to \infty$; $\mu(\varepsilon_j)$ are positive roots of the equations $4|A^{[j]}_{\mu}(b_3)| = M_1\varepsilon_j^{-1}$.

Theorem 2.3.5. *If the function $u_1(x,y)$ satisfies conditions (2.3.55)–(2.3.60) and*

$$\|\Delta u_1(x,b_3)\|_{L_2(0,a_3)} \le M, \tag{2.3.61}$$

then the inequality

$$\|u_1(x,y)\|_{L_2(0,a_3)} \le \frac{M}{4}\sum_{j=1}^{4}\left|\frac{A^{[j]}_{\bar\mu(\varepsilon_j)}(y)}{B^{[j]}_{\bar\mu(\varepsilon_j)}(b_3)}\right|$$

holds. Here $A_{\bar{\mu}}^{[j]}(y)$ and $B_{\bar{\mu}}^{[j]}(y)$ are known functions, and

$$\left|\frac{A_{\bar{\mu}}^{[j]}(y)}{B_{\bar{\mu}}^{[j]}(b_3)}\right| \to 0$$

as $\bar{\mu} \to \infty$; $\bar{\mu}(\varepsilon_j)$ are positive roots of the equations $4|B_{\bar{\mu}}^{[j]}(b_3)| = M\varepsilon_j^{-1}$.

Suppose that $\varphi_j(x)$, $j = \overline{1,4}$, are known with accuracy δ; i.e., we know $\varphi_{j\delta}(x)$ such that

$$\|\varphi_j(x) - \varphi_{j\delta}(x)\|_{L_2} \leq \delta.$$

The approximate solution of the problem (2.3.22)–(2.3.26) can be represented as follows

$$u_{1N\delta}(x,y) = \sum_{k=1}^{4} u_{1kN\delta}(x,y),$$

where $u_{1kN\delta}(x,y)$ are approximate solutions of the problems (2.3.32)–(2.3.36).

The following estimates hold

$$\|u_{1N\delta}(x,y) - u_1(x,y)\|_{L_2(0,a_3)} \leq \omega_1(N,\delta,y), \tag{2.3.62}$$

$$\|u_{1N\delta}(x,y) - u_1(x,y)\|_{L_2(0,a_3)} \leq \omega_2(N,\delta,y), \tag{2.3.63}$$

where

$$\omega_1(N,\delta,y) = \sum_{k=1}^{4}\left\{|A_N^{[k]}(y)|\delta + \frac{M_1}{4}\left|\frac{A_{N+1}^{[k]}(y)}{A_{N+1}^{[k]}(b_3)}\right|\right\},$$

$$\omega_2(N,\delta,y) = \sum_{k=1}^{4}\left\{|A_N^{[k]}(y)|\delta + \frac{M}{4}\left|\frac{A_{N+1}^{[k]}(y)}{B_{N+1}^{[k]}(b_3)}\right|\right\}.$$

If the accuracy δ is fixed, then the value of N when $\inf \omega_1(N,\delta,y)$ ($\inf \omega_2(N,\delta,y)$) is achieved will be optimal in the sense of the estimate (2.3.62) (or (2.3.63)).

Note that analogous results hold for the problem (2.3.27)–(2.3.31).

Suppose now that $a_1 = b_1$, $a_2 = b_2$, $a_3 = b_3 = a$, $0 < a_1 < a_2 < a < \infty$.

We suppose that the functions $\psi_j(y)$, $\varphi_j(x)$ are known approximately; i.e., instead of them we know their approximations $\psi_{j\delta}(y)$, $\varphi_{j\delta}(x)$ such that

$$\|\psi_j(y) - \psi_{j\delta}(y)\|_{L_2} \leq \delta, \qquad \|\varphi_j(x) - \varphi_{j\delta}(x)\|_{L_2} \leq \delta, \qquad j = 1,2,3,4.$$

As an approximate solution of (2.3.1)–(2.3.5) we take the function

$$u_\delta^\alpha(x,y) = \sum_{m=0}^{\infty}\Big\{\big[\alpha_m A_m(y) + \beta_m B_m(y) + \bar\alpha_m \bar A_m(y) + \bar\beta_m \bar B_m(y)\big]\cos\frac{m\pi x}{a} + \big[c_m A_m(x) + d_m B_m(x) + \bar c_m \bar A_m(x) + \bar d_m \bar B_m(x)\big]\cos\frac{m\pi y}{a}\Big\}$$

where

$$A_0(t) = 1 - \frac{4}{a^2}t^2 + \frac{2}{a^3}t^3, \qquad B_0(t) = \frac{3}{a^2}t^2 - \frac{2}{a^3}t^3,$$

$$\bar A_0(t) = t - \frac{2}{a^2}t^2 + \frac{1}{a^2}t^3, \qquad \bar B_0(t) = \frac{1}{a^3}t^3 - \frac{1}{a^3}t^2;$$

$A_m(t)$ and $B_m(t)$ are defined by formulas (2.3.8) and (2.3.9) for $a_3 = b_3 = a$;

$$\bar A_m(t) = \frac{m\pi(a-t)\sinh(m\pi t/a) - t\sinh(m\pi)\sinh(m\pi(a-t)/a)}{\sinh^2(m\pi) - m^2\pi^2},$$

$$\bar B_m(t) = \frac{(a-t)\sinh(m\pi)\sinh(m\pi t/a) - m\pi t\sinh(m\pi(a-t)/a)}{\sinh^2(m\pi) - m^2\pi^2};$$

α is the regularization parameter; α_m, β_m, c_m, d_m, $\bar\alpha_m$, $\bar\beta_m$, $\bar c_m$, and $\bar d_m$ are the Fourier coefficients of the functions $\Phi_{1\delta}^{[\alpha]}(x)$, $\Phi_{2\delta}^{[\alpha]}(x)$, $\Phi_{3\delta}^{[\alpha]}(x)$, $\Phi_{4\delta}^{[\alpha]}(x)$, $\varphi_{1\delta}(x)$, $\varphi_{2\delta}(x)$, $\psi_{3\delta}(x)$, and $\psi_{4\delta}(x)$. The functions $\Phi_{k\delta}^{[\alpha]}(x)$, $k = 1,2,3,4$, are an approximate solution of the system of integral equations of first kind

$$\sum_{k=1}^{4}\int_0^a K_{jk}(x,s)\Phi_k(s)\,\mathrm{d}s = f_{j\delta}(x), \qquad a_1 \le x \le a_2, \quad j = 1,2,3,4. \tag{2.3.64}$$

Here (in the formulas below, $n = 1,2$)

$$K_{n1}(x,s) = \frac{2}{a}\sum_{m=0}^{\infty} A_m(a_n)\cos\frac{m\pi x}{a}\cos\frac{m\pi s}{a},$$

$$K_{n2}(x,s) = \frac{2}{a}\sum_{m=0}^{\infty} B_m(a_n)\cos\frac{m\pi x}{a}\cos\frac{m\pi s}{a},$$

$$K_{n3}(x,s) = \frac{2}{a}\sum_{m=0}^{\infty} A_m(x)\cos\frac{m\pi a_n}{a}\cos\frac{m\pi s}{a},$$

$$K_{n4}(x,s) = \frac{2}{a}\sum_{m=0}^{\infty} B_m(x)\cos\frac{m\pi a_n}{a}\cos\frac{m\pi s}{a},$$

$$K_{2+n,1}(x,s) = K_{n3}(x,s), \qquad K_{2+n,3}(x,s) = K_{n1}(x,s),$$
$$K_{2+n,2}(x,s) = K_{n4}(x,s), \qquad K_{2+n,4}(x,s) = K_{n2}(x,s),$$

$$f_{n\delta}(x) = \varphi_{2+n,\delta}(x) - \sum_{m=0}^{\infty}\Big\{\big[\bar{\alpha}_m\bar{A}_m(a_n) + \bar{\beta}_m\bar{B}_m(a_n)\big]\cos\frac{m\pi x}{a}$$
$$+ \big[\bar{c}_m\bar{A}_m(x) + \bar{d}_m\bar{B}_m(x)\big]\cos\frac{m\pi a_n}{a}\Big\},$$

$$f_{2+n,\delta}(x) = \psi_{2+n,\delta}(x) - \sum_{m=0}^{\infty}\Big\{\big[\bar{\alpha}_m\bar{A}_m(x) + \bar{\beta}_m\bar{B}_m(x)\big]\cos\frac{m\pi x}{a}$$
$$+ \big[\bar{c}_m\bar{A}_m(a_n) + \bar{d}_m\bar{B}_m(a_n)\big]\cos\frac{m\pi x}{a}\Big\}.$$

Let us transform system (2.3.64) into one equation. To this end, we set

$$f_\delta(x) = \begin{cases} f_{1\delta}(x), & a_1 \le x \le a_2, \\ f_{2\delta}(x + a_1 - a_2), & a_2 < x \le 2a_2 - a_1, \\ f_{3\delta}(x + 2a_1 - 2a_2), & 2a_2 - a_1 < x \le 3a_2 - 2a_1, \\ f_{4\delta}(x + 3a_1 - 3a_2), & 3a_2 - 2a_1 < x \le 4a_2 - 3a_1, \end{cases}$$

$$\Phi(s) = \begin{cases} \Phi_1(s), & 0 \le s \le a, \\ \Phi_2(s-a), & a < s \le 2a, \\ \Phi_3(s-2a), & 2a < s \le 3a, \\ \Phi_4(s-3a), & 3a < s \le 4a, \end{cases}$$

$$K(x,s) = \begin{cases} K_{11}(x,s), & a_1 \le x \le a_2,\ 0 \le s \le a, \\ K_{1,j+1}(x, s-ja), & a_1 \le x \le a_2, \\ & ja < s \le (j+1)a,\ j = 1,2,3, \\ K_{21}(x+a_1-a_2, s), & a_2 < x \le 2a_2 - a_1,\ 0 \le s \le a, \\ K_{2,j+1}(x+a_1-a_2, s-ja), & 2a_2 - a_1 < x \le 3a_2 - 2a_1, \\ & ja < s \le (j+1)a,\ j = 1,2,3, \\ K_{31}(x+2a_1-2a_2, s), & 2a_2 - a_1 < x \le 3a_2 - 2a_1, \\ & 0 \le s \le a, \\ K_{3,j+1}(x+2a_1-2a_2, s-ja), & 2a_2 - a_1 < x \le 3a_2 - 2a_1, \\ & ja < s \le (j+1)a,\ j = 1,2,3, \\ K_{41}(x+3a_1-3a_2, s), & 3a_2 - 2a_1 < x \le 4a_2 - 3a_1, \\ & 0 \le s \le a, \\ K_{4,j+1}(x+3a_1-3a_2, s-ja), & 3a_2 - 2a_1 < x \le 4a_2 - 3a_1, \\ & ja < s \le (j+1)a,\ j = 1,2,3. \end{cases}$$

Now system (2.3.64) may be written as one equation

$$\int_0^{4a} K(x,s)\Phi(s)\,\mathrm{d}s = f_\delta(x), \qquad a_1 \le x \le 4a_2 - 3a_1,$$

whose approximate solution is constructed by the Tikhonov regularization method (see Tikhonov and Arsenin, 1977).

The following estimate holds

$$\|u(x,y) - u_\delta^\alpha(x,y)\|_{C(D_1)} \le \sum_{k=1}^{4} h_k \omega_k(\delta) + h_5 \delta,$$

where

$$\omega_k(\delta) \ge \|\Phi_k(s) - \Phi_{k\delta}^{[\alpha]}(s)\|_{C(0,a)}, \qquad k = 1,2,3,4,$$

and $\omega_k(\delta) \to 0$ for $\delta \to 0$;

$$h_k = \max_{(x,y)\in D_1} \left\{ \int_0^a |K_k(x,y,s)|\,\mathrm{d}s \right\}, \qquad k = 1,2,3,4,$$

$$h_5 = \sum_{k=5}^{8} \left\{ \max_{(x,y)\in D_1} \left[\int_0^a K_k^2(x,y,s)\,\mathrm{d}s \right]^{1/2} \right\}.$$

In the above formulas we used the notation

$$K_1(x,y,s) = \frac{2}{a}\sum_{m=0}^{\infty} A_m(y)\cos\frac{m\pi x}{a}\cos\frac{m\pi s}{a},$$

$$K_2(x,y,s) = \frac{2}{a}\sum_{m=0}^{\infty} B_m(y)\cos\frac{m\pi x}{a}\cos\frac{m\pi s}{a},$$

$$K_3(x,y,s) = K_1(x,y,s), \qquad K_4(x,y,s) = K_2(x,y,s),$$

$$K_5(x,y,s) = \frac{2}{a}\sum_{m=0}^{\infty} \bar{A}_m(x)\cos\frac{m\pi y}{a}\cos\frac{m\pi s}{a},$$

$$K_6(x,y,s) = \frac{2}{a}\sum_{m=0}^{\infty} \bar{B}_m(x)\cos\frac{m\pi y}{a}\cos\frac{m\pi s}{a},$$

$$K_7(x,y,s) = K_5(y,x,s), \qquad K_8(x,y,s) = K_6(y,x,s).$$

Note that if $a_1 \neq b_1$, $a_2 \neq b_2$, and $a_3 \neq b_3$, then, extending by zero in bigger segments the corresponding functions from the smaller ones, we arrive at the case considered above.

Case B. Let

$$D_1 = \{(x,y) \mid 0 < x < b,\ 0 < y < b\}, \qquad S_1 = \partial D_1,$$

$$D_2 = \{(x,y) \mid a_1 < x < a_2,\ a_1 < y < x\}, \qquad S_2 = \partial D_2.$$

Here $0 < a_1 < a_2 < b < \infty$; and ν will denote the external normal to S_1.

Problem 2.3.2. We need to find the function $u(x,y)$ in D_1 satisfying the conditions

$$\Delta^2 u(x,y) = 0 \quad \text{in} \quad D_1, \tag{2.3.65}$$

$$u'_x(0,y) = \psi_1(y), \qquad u'_x(b,y) = \psi_2(y), \qquad 0 \le y \le b, \tag{2.3.66}$$

$$u'_y(x,0) = \varphi_1(x), \qquad u'_y(x,b) = \varphi_2(x), \qquad 0 \le x \le b, \tag{2.3.67}$$

$$u(x,a_1) = \varphi_3(x), \qquad u(x,x) = f(x), \qquad a_1 \le x \le a_2, \tag{2.3.68}$$

$$u(a_2,y) = \psi_3(y), \qquad a_1 \le y \le a_2. \tag{2.3.69}$$

Here $f(x)$, $\psi_j(y)$, and $\varphi_j(x)$, $j = 1,2,3$, are given functions such that

$$\psi_3(a_2) = f(a_2), \qquad \varphi_3(a_1) = f(a_1), \qquad \psi_3(a_1) = \varphi_3(a_2).$$

Let us show that in this problem there is no continuous dependence of the solution on the data in any pair of spaces. Really, the function

$$u_m(x,y) = \varepsilon\Big[\widetilde{C}_m\big(\tfrac{3\pi}{2}\big)\widetilde{D}_m\big(\tfrac{\pi}{2}\big) - \widetilde{C}_m\big(\tfrac{\pi}{2}\big)\widetilde{D}_m\big(\tfrac{3\pi}{2}\big)\Big]^{-1}$$
$$\times\Big\{\Big[\widetilde{C}_m(x)\widetilde{D}_m\big(\tfrac{3\pi}{2}\big) - \widetilde{C}_m\big(\tfrac{3\pi}{2}\big)\widetilde{D}_m(x)\Big]\cos(2m+1)y$$
$$-\Big[\widetilde{C}_m(y)\widetilde{D}_m\big(\tfrac{3\pi}{2}\big) - \widetilde{C}_m\big(\tfrac{3\pi}{2}\big)\widetilde{D}_m(y)\Big]\cos(2m+1)x\Big\}, \tag{2.3.70}$$

where

$$\widetilde{C}_m(t) = \sinh 2M\pi\big[\sinh M(2\pi - t) + Mt\cosh M(2\pi - t)\big]$$
$$- 2M\pi\big[\sinh Mt + M(2\pi - t)\cosh Mt\big],$$
$$\widetilde{D}_m(t) = \sinh 2M\pi\,\sinh Mt + M\big[2\pi\cosh 2M\pi\,\sinh Mt$$
$$- t\sinh 2M\pi\,\cosh Mt\big] - 2M^2\pi t\cosh M(2\pi - t),$$
$$M = 2m+1,$$

is the solution of the problem (2.3.65)–(2.3.69) for $\psi_j(y) = 0$, $\varphi_j(x) = 0$, $j = 1,2$, $f(x) = 0$, $\psi_3(y) = 0$, $\varphi_3(x) = \varepsilon\cos(2m+1)x$, $b = 2\pi$, $a_1 = \pi/2$, and $a_2 = 3\pi/2$.

From (2.3.70) it follows that for each pair of functional spaces of functions depending on x and for any constants $\bar{\varepsilon} > 0$ and $C > 0$, we may choose ε and m such that

$$\|\varepsilon\cos(2m+1)x\| < \bar{\varepsilon}, \qquad \|u_m(x,y)\| > C.$$

Now let us consider the uniqueness of solution of (2.3.65)–(2.3.69). For simplicity, we set

$$\psi_1(y) = \psi_2(y) = 0, \qquad \varphi_1(x) = \varphi_2(x) = 0.$$

A function $u(x,y)$ biharmonic in D_1 and satisfying the condition

$$\frac{\partial u}{\partial \nu}\Big|_{S_1} = 0$$

can be written in the form (see formulas (2.3.7)–(2.3.11) where we set $a_3 = b_3 = b$)

$$u(x,y) = \alpha_0 + \beta_0 D_0(y) + c_0 D_0(x)$$

$$+\sum_{m=1}^{\infty}\Big\{\big[\alpha_m C_m(y)+\beta_m D_m(y)\big]\cos\frac{m\pi x}{b}$$
$$+\big[c_m C_m(x)+d_m D_m(x)\big]\cos\frac{m\pi y}{b}\Big\}, \tag{2.3.71}$$

where

$$D_0(t)=\frac{2}{b^3}\,t^3-\frac{3}{b^2}\,t^2, \tag{2.3.72}$$

$$C_m(t)=\Big[\sinh m\pi\Big(\sinh\frac{m\pi(b-t)}{b}+\frac{m\pi t}{b}\cosh\frac{m\pi(b-t)}{b}\Big)$$
$$-m\pi\Big(\sinh\frac{m\pi t}{b}+\frac{m\pi(b-t)}{b}\cosh\frac{m\pi t}{b}\Big)\Big]\Theta, \tag{2.3.73}$$

$$D_m(t)=\Big[\sinh m\pi\,\sinh\frac{m\pi t}{b}-\frac{m^2\pi^2 t}{b}\cosh\frac{m\pi(b-t)}{b}$$
$$+\frac{m\pi}{b}\Big(b\cosh m\pi\,\sinh\frac{m\pi t}{b}-t\sinh m\pi\,\cosh\frac{m\pi t}{b}\Big)\Big]\Theta, \tag{2.3.74}$$

$$\Theta=\frac{1}{\sinh^2 m\pi-m^2\pi^2},$$

α_m, β_m, c_m, and d_m, $m=1,2,3,\dots$, are arbitrary constants.

The following statement holds.

Theorem 2.3.6. *If*

$$|\alpha_m|,\ |\beta_m|,\ |c_m|,\ |d_m|\le C, \tag{2.3.75}$$

where C is some number, then the equalities

$$u(a_2,y)=0,\qquad a_1\le y\le a_2, \tag{2.3.76}$$
$$u(x,a_1)=u(x,x)=0,\qquad a_1\le x\le a_2, \tag{2.3.77}$$

imply that $u(x,y)=0$ in D_1.

Proof. The proof of this theorem is similar to that of Theorem 2.3.1, however, there are some peculiarities.

From (2.3.76), (2.3.77) it follows that the functions

$$\psi_3(y)=\alpha_0+\beta_0 D_0(y)+c_0 D_0(a_2)$$
$$+\sum_{m=1}^{\infty}\Big\{\big[\alpha_m C_m(y)+\beta_m D_m(y)\big]\cos\frac{m\pi a_2}{b}$$

$$+ \left[c_m C_m(a_2) + d_m D_m(a_2)\right] \cos \frac{m\pi y}{b} \Big\},$$

$$\varphi_3(x) = \alpha_0 + \beta_0 D_0(a_1) + c_0 D_0(x) + \sum_{m=1}^{\infty} \Big\{ \left[\alpha_m C_m(a_1) + \beta_m D_m(a_1)\right] \cos \frac{m\pi z}{b} + \left[c_m C_m(x) + d_m D_m(x)\right] \cos \frac{m\pi a_1}{b} \Big\},$$

$$f(x) = \alpha_0 + (\beta_0 + c_0) D_0(x) + \sum_{m=1}^{\infty} \Big\{ [\alpha_m + c_m] C_m(x) \cos \frac{m\pi x}{b} + [\beta_m + d_m] D_m(x) \cos \frac{m\pi x}{b} \Big\},$$

vanish on the segment $[a_1, a_2]$.

From (2.3.75) it follows that the functions $\psi_3(t)$, $\varphi_3(t)$, and $f(t)$ can be analytically continued onto the rectangle

$$\overline{P} = \left\{ z = t + \mathrm{i}\xi \;\middle|\; -2b \le \operatorname{Re} z, \operatorname{Im} z \le 2b \right\}$$

by means of the equalities

$$\psi_3(z) = a_{1,0}^{[0]} + a_{1,2}^{[0]} z^2 + a_{1,3}^{[0]} z^3 + \sum_{m=1}^{\infty} \Big[a_{m,0}^{[1]} \mathrm{e}^{(m\pi/b)z} + a_{m,1}^{[1]} z \mathrm{e}^{(m\pi/b)z} + a_{m,0}^{[2]} \mathrm{e}^{-(m\pi/b)z} + a_{m,1}^{[2]} z \mathrm{e}^{-(m\pi/b)z} + a_{m,0}^{[3]} \mathrm{e}^{\mathrm{i}(m\pi/b)z} + a_{m,0}^{[4]} \mathrm{e}^{-\mathrm{i}(m\pi/b)z} \Big], \tag{2.3.78}$$

$$\varphi_3(z) = b_{1,0}^{[0]} + b_{1,2}^{[0]} z^2 + b_{1,3}^{[0]} z^3 + \sum_{m=1}^{\infty} \Big[b_{m,0}^{[1]} \mathrm{e}^{(m\pi/b)z} + b_{m,1}^{[1]} z \mathrm{e}^{(m\pi/b)z} + b_{m,0}^{[2]} \mathrm{e}^{-(m\pi/b)z} + b_{m,1}^{[2]} z \mathrm{e}^{-(m\pi/b)z} + b_{m,0}^{[3]} \mathrm{e}^{\mathrm{i}(m\pi/b)z} + b_{m,0}^{[4]} \mathrm{e}^{-\mathrm{i}(m\pi/b)z} \Big], \tag{2.3.79}$$

$$f(z) = C_{1,0}^{[0]} + C_{1,2}^{[0]} z^2 + C_{1,3}^{[0]} z^3 + \sum_{m=1}^{\infty} \Big[C_{m,0}^{[1]} \mathrm{e}^{(1+\mathrm{i})(m\pi/b)z} + C_{m,1}^{[1]} z \mathrm{e}^{(1+\mathrm{i})(m\pi/b)z} + C_{m,0}^{[2]} \mathrm{e}^{(1-\mathrm{i})(m\pi/b)z} + C_{m,1}^{[2]} z \mathrm{e}^{(1-\mathrm{i})(m\pi/b)z} + C_{m,0}^{[3]} \mathrm{e}^{(-1-\mathrm{i})(m\pi/b)z} + C_{m,1}^{[3]} z \mathrm{e}^{(-1-\mathrm{i})(m\pi/b)z} + C_{m,0}^{[4]} \mathrm{e}^{(-1+\mathrm{i})(m\pi/b)z} + C_{m,1}^{[4]} z \mathrm{e}^{(-1+\mathrm{i})(m\pi/b)z} \Big], \tag{2.3.80}$$

where

$$a_{1,0}^{[0]} = \alpha_0 + c_0 D_0(a_2), \qquad a_{1,2}^{[0]} = -\frac{3\beta_0}{b^2}, \qquad a_{1,3}^{[0]} = -\frac{2\beta_0}{b^3},$$

$$\begin{aligned}
a_{m,0}^{[1]} &= \bar{K}_m \cos\frac{m\pi a_2}{b}\big[-\alpha_m(\sinh m\pi \cdot \mathrm{e}^{-m\pi} + m\pi(1+m\pi)) \\
&\qquad + \beta_m(\sinh m\pi + m\pi\cosh m\pi)\big], \\
a_{m,1}^{[1]} &= \bar{K}_m \cos\frac{m\pi a_2}{b}\Big[\alpha_m(\sinh m\pi \cdot \mathrm{e}^{-m\pi} + m\pi)\frac{m\pi}{b} \\
&\qquad - \beta_m(\sinh m\pi + m\pi\mathrm{e}^{-m\pi})\frac{m\pi}{b}\Big], \\
a_{m,0}^{[2]} &= \bar{K}_m \cos\frac{m\pi a_2}{b}\big[\alpha_m(\sinh m\pi \cdot \mathrm{e}^{-m\pi} + m\pi(1-m\pi)) \\
&\qquad - \beta_m(\sinh m\pi + m\pi\cosh m\pi)\big], \\
a_{m,1}^{[2]} &= \bar{K}_m \cos\frac{m\pi a_2}{b}\Big[\alpha_m(\sinh m\pi \cdot \mathrm{e}^{m\pi} + m\pi)\frac{m\pi}{b} \\
&\qquad - \beta_m(\sinh m\pi + m\pi\mathrm{e}^{m\pi})\frac{m\pi}{b}\Big],
\end{aligned}$$

$$a_{m,0}^{[3]} = a_{m,0}^{[4]} = \frac{c_m C_m(a_2) + \alpha_m D_m(a_2)}{2},$$

$$b_{1,0}^{[0]} = \alpha_0 + \beta_0 D_0(a_1), \qquad b_{1,2}^{[0]} = -\frac{3c_0}{b^2}, \qquad b_{1,3}^{[0]} = \frac{2c_0}{b^3},$$

$$\begin{aligned}
b_{m,0}^{[1]} &= \bar{K}_m \cos\frac{m\pi a_1}{b}\big[-c_m(\sinh m\pi \cdot \mathrm{e}^{-m\pi} + m\pi(1-m\pi)) \\
&\qquad + d_m(\sinh m\pi + m\pi\cosh m\pi)\big], \\
b_{m,1}^{[1]} &= \bar{K}_m \cos\frac{m\pi a_1}{b}\Big[c_m(\sinh m\pi \cdot \mathrm{e}^{-m\pi} + m\pi)\frac{m\pi}{b} \\
&\qquad - d_m(\sinh m\pi + m\pi\mathrm{e}^{-m\pi})\frac{m\pi}{b}\Big], \\
b_{m,0}^{[2]} &= \bar{K}_m \cos\frac{m\pi a_1}{b}\big[c_m(\sinh m\pi \cdot \mathrm{e}^{m\pi} + m\pi(1-m\pi)) \\
&\qquad - d_m(\sinh m\pi + m\pi\cosh m\pi)\big], \\
b_{m,1}^{[2]} &= \bar{K}_m \cos\frac{m\pi a_1}{b}\Big[c_m(\sinh m\pi \cdot \mathrm{e}^{-m\pi} + m\pi)\frac{m\pi}{b} \\
&\qquad - d_m(\sinh m\pi + m\pi\mathrm{e}^{-m\pi})\frac{m\pi}{b}\Big],
\end{aligned}$$

$$b_{m,0}^{[3]} = b_{m,0}^{[4]} = \frac{\alpha_m C_m(a_1) + \beta_m D_m(a_1)}{2},$$

$$C_{1,0}^{[0]} = \alpha_0, \qquad C_{1,2}^{[0]} = -\frac{3(\beta_0 + c_0)}{b^2}, \qquad C_{1,3}^{[0]} = \frac{2(\beta_0 + c_0)}{b^3},$$

$$C_{m,0}^{[1]} = C_{m,0}^{[2]} = \frac{\bar{K}_m}{2}\big[-(\alpha_m + c_m)(\sinh m\pi \cdot \mathrm{e}^{-m\pi} + m\pi(1 + m\pi)) \\ + (\beta_m + d_m)(\sinh m\pi + m\pi \cosh m\pi)\big],$$

$$C_{m,1}^{[1]} = C_{m,1}^{[2]} = \frac{\bar{K}_m}{2}\Big[(\alpha_m + c_m)(\sinh m\pi \cdot \mathrm{e}^{-m\pi} + m\pi)\frac{m\pi}{b} \\ - (\beta_m + d_m)(\sinh m\pi + m\pi \mathrm{e}^{-m\pi})\frac{m\pi}{b}\Big],$$

$$C_{m,0}^{[3]} = C_{m,0}^{[4]} = \frac{\bar{K}_m}{2}\big[(\alpha_m + c_m)(\sinh m\pi \cdot \mathrm{e}^{m\pi} + m\pi(1 - m\pi)) \\ - (\beta_m + d_m)(\sinh m\pi + m\pi \cosh m\pi)\big],$$

$$C_{m,1}^{[3]} = C_{m,1}^{[4]} = \frac{\bar{K}_m}{2}\Big[(\alpha_m + c_m)(\sinh m\pi \cdot \mathrm{e}^{m\pi} + m\pi)\frac{m\pi}{b} \\ - (\beta_m + d_m)(\sinh m\pi + m\pi \mathrm{e}^{m\pi})\frac{m\pi}{b}\Big],$$

$$\bar{K}_m = \frac{1}{2(\sinh^2 m\pi - m^2\pi^2)}.$$

Since $\psi_3(z) = 0$, $\varphi_3(z) = 0$, $f(z) = 0$ on the real segment $[a_1, a_2]$, therefore, from uniqueness of continuation of analytic functions of one complex variable we conclude that $\psi_3(z) = 0$, $\varphi_3(z) = 0$, $f(z) = 0$ everywhere in the rectangle $\overline{P}$.

We introduce now the entire function of exponential type

$$Q(\lambda) = \lambda \sinh \lambda b \sin^2 \lambda b = \frac{1}{2}\sum_{n=0}^{\infty}\big[1 - 5^{n+1/2}\cos(2n+1)\bar{\varphi}_0\big]\frac{b^{2n+1}\lambda^{2n+2}}{(2n+1)!},$$

where $\bar{\varphi}_0 = \arctan 2$.

At the points

$$\lambda_1^{[0]} = 0, \quad \lambda_m^{[1]} = \frac{m\pi}{b}, \quad \lambda_m^{[2]} = -\frac{m\pi}{b}, \quad \lambda_m^{[3]} = \frac{\mathrm{i}m\pi}{b}, \quad \lambda_m^{[4]} = -\frac{\mathrm{i}m\pi}{b}$$

$(m = 1, 2, \dots)$, the function $Q(\lambda)$ has zeros of the orders

$$\nu_1^{[0]} = 4, \qquad \nu_m^{[1]} = \nu_m^{[2]} = 2, \qquad \nu_m^{[3]} = \nu_m^{[4]} = 1, \qquad m = 1, 2, \dots,$$

respectively.

We will denote by $\gamma(z)$ the Borel functional transform of $Q(\lambda)$. Then

$$\gamma(z)=\frac{1}{2}\sum_{n=0}^{\infty}\left[1-5^{n+1/2}\cos(2n+1)\bar{\varphi}_0\right]\frac{b^{2n+1}(2n+2)}{z^{2n+3}}$$
$$=\frac{bz}{(z^2-b^2)^2}-\frac{\sqrt{5}\,bz\left[z^4-10b^2z^2+25b^4(2\cos 2\bar{\varphi}_0-1)\right]^2}{(z^4-10b^2z^2\cos 2\bar{\varphi}_0+25b^4)^2}\cos\bar{\varphi}_0.$$

The function $\gamma(z)$ has singularities only at the points

$$z=b,\quad -b,\quad (1+2\mathrm{i})b,\quad (1-2\mathrm{i})b,\quad (-1+2\mathrm{i})b,\quad (-1-2\mathrm{i})b.$$

The smallest convex domain which contains all these singularities of $\gamma(z)$ is the rectangle

$$\overline{P_1}=\left\{z\;\middle|\;-b\le \operatorname{Re} z\le b,\ -2b\le \operatorname{Im} z\le 2b\right\}.$$

By condition (2.3.75) of the theorem, we have $\overline{P_1}\subset P$.

Let Γ_1 be a simple closed contour lying in $P\setminus\overline{P_1}$.

We will denote by $D^{[i]}_{m,k}(z)$ the Borel functional transform of the entire function $Q(\lambda)/(\lambda-\lambda^{[i]}_m)^{k+1}$; $k=0,1,2,\dots,\nu^{[i]}_m-1$; $i=0,1,2,3,4$. Let

$$M^{[i]}_m(\mu)=\frac{(\mu-\lambda^{[i]}_m)^{\nu^{[i]}_m}}{Q(\mu)},$$

$$\Phi^{[i]}_{m,k}(z)=\frac{1}{k!\,(\nu^{[i]}_m-k-1)!}\sum_{q=0}^{\nu^{[i]}_m-k-1}q!\cdot C^q_{\nu^{[i]}_m-k-1}\left[M^{[i]}_m(\mu)\right]^{(\nu^{[i]}_m-k-q-1)}_{\mu=\lambda^{[i]}_m}D^{[i]}_{m,q}(z).$$

Then

$$\frac{1}{2\pi\mathrm{i}}\int_{\Gamma_1}z^s\mathrm{e}^{\lambda^{[i]}_p z}\Phi^{[n]}_{m,k}(z)\,\mathrm{d}z=\begin{cases}1, & i=n,\ p=m,\ s=k,\\ 0, & \text{otherwise.}\end{cases}\tag{2.3.81}$$

Multiply both sides of equalities (2.3.78) and (2.3.79) by the functions

$$\frac{1}{2\pi\mathrm{i}}\Phi^{[0]}_{1,0}(z),\qquad \frac{1}{2\pi\mathrm{i}}\Phi^{[3]}_{m,0}(z),\qquad m=1,2,\dots,$$

and integrate around the contour Γ_1. Then, by (2.3.81), we obtain

$$a^{[0]}_{1,0}=\int_{\Gamma_1}\psi_3(z)\Phi^{[0]}_{1,0}(z)\,\mathrm{d}z,\qquad a^{[3]}_{m,0}=\int_{\Gamma_1}\psi_3(z)\Phi^{[3]}_{m,0}(z)\,\mathrm{d}z,$$

$$b^{[0]}_{1,0} = \int_{\Gamma_1} \varphi_3(z)\Phi^{[0]}_{1,0}(z)\,\mathrm{d}z, \qquad b^{[3]}_{m,0} = \int_{\Gamma_1} \varphi_3(z)\Phi^{[3]}_{m,0}(z)\,\mathrm{d}z. \tag{2.3.82}$$

Now consider the function

$$\begin{aligned}\widetilde{Q}(\lambda) &= \sin^2\frac{\lambda b}{1+\mathrm{i}}\cdot\sin^2\frac{\lambda b}{1-\mathrm{i}}\\ &= \frac{1}{4}\sum_{n=1}^{\infty}\frac{(2b)^{4n}\lambda^{4n}}{(4n)!} - \frac{1}{2}\sum_{n=1}^{\infty}(-1)^n\frac{(2b)^{4n}\lambda^{4n}}{(4n)!};\end{aligned}$$

$\widetilde{Q}(\lambda)$ is an entire function of exponential type, which at the points

$$\tilde{\lambda}^{[0]}_1 = 0, \qquad \tilde{\lambda}^{[1]}_m = \frac{(1+\mathrm{i})m\pi}{b}, \qquad \tilde{\lambda}^{[2]}_m = \frac{(1-\mathrm{i})m\pi}{b},$$

$$\tilde{\lambda}^{[3]}_m = -\frac{(1+\mathrm{i})m\pi}{b}, \qquad \tilde{\lambda}^{[4]}_m = -\frac{(1-\mathrm{i})m\pi}{b}, \qquad m = 1, 2, \ldots,$$

has zeros of the orders

$$\tilde{\nu}^{[0]}_1 = 4, \qquad \tilde{\nu}^{[1]}_m = \tilde{\nu}^{[2]}_m = 2, \qquad \tilde{\nu}^{[3]}_m = \tilde{\nu}^{[4]}_m = 2, \qquad m = 1, 2, \ldots,$$

respectively.

We will denote by $\tilde{\gamma}(z)$ the Borel functional transform of $\widetilde{Q}(\lambda)$. Then

$$\begin{aligned}\tilde{\gamma}(z) &= \frac{1}{4}\sum_{n=1}^{\infty}\frac{(2b)^{4n}}{z^{4n+1}} - \frac{1}{2}\sum_{n=1}^{\infty}(-1)^n\frac{(\sqrt{2}\,b)^{4n}}{z^{4n+1}}\\ &= \frac{1}{4z} + \frac{1}{4}\frac{z^3}{z^4-16b^4} - \frac{1}{2}\frac{z^3}{z^4+4b^4} = \frac{2b^4(3z^4-8b^4)}{z(z^4-16b^4)(z^4+4b^4)}.\end{aligned}$$

This function has singularities only at the points

$$z = 0,\ 2b,\ -2b,\ 2b\mathrm{i},\ -2b\mathrm{i},\ (1+\mathrm{i})b,\ -(1+\mathrm{i})b,\ (1-\mathrm{i})b,\ -(1-\mathrm{i})b.$$

The smallest convex domain which contains all these singularities of $\tilde{\gamma}(z)$ is the square $\overline{P_2}$ with the vertices $z = 2b, -2b, 2b\mathrm{i}, -2b\mathrm{i}$. Condition (2.3.75) of the theorem implies that $\overline{P_2} \subset P$.

Suppose now that Γ_2 is a simple closed contour lying in $P \setminus \overline{P_2}$.

We will denote by $\widetilde{D}^{[i]}_{m,k}$ the Borel functional transform of the entire function $Q(\lambda)/(\lambda - \tilde{\lambda}^{[i]}_m)^{k+1}$; $k = 0, 1, 2, \ldots, \tilde{\nu}^{[i]}_m - 1$; $i = 0, 1, 2, 3, 4$. Let

$$\widetilde{M}^{[i]}_m(\mu) = \frac{(\mu - \tilde{\lambda}^{[i]}_m)^{\tilde{\nu}^{[i]}_m}}{\widetilde{Q}(\mu)},$$

$$f_{m,k}^{[i]}(z) = \frac{1}{k!\,(\tilde{\nu}_m^{[i]} - k - 1)!} \sum_{q=0}^{\tilde{\nu}_m^{[i]}-k-1} q! \cdot C_{\tilde{\nu}_m^{[i]}-k-1}^{q} [\widetilde{M}_m^{[i]}(\mu)]_{\mu=\tilde{\lambda}_m^{[i]}}^{(\tilde{\nu}_m^{[i]}-k-q-1)} \widetilde{D}_{m,q}^{[i]}(z).$$

Then

$$\frac{1}{2\pi \mathrm{i}} \int_{\Gamma_2} z^s \mathrm{e}^{\tilde{\lambda}_p^{[i]} z} f_{m,k}^{[n]}(z)\,\mathrm{d}z = \begin{cases} 1, & i = n,\ p = m,\ s = k, \\ 0, & \text{otherwise.} \end{cases} \tag{2.3.83}$$

Multiply both sides of equality (2.3.80) by the functions

$$\frac{1}{2\pi \mathrm{i}} f_{1,0}^{[0]}(z), \qquad \frac{1}{2\pi \mathrm{i}} f_{m,0}^{[1]}(z), \qquad \frac{1}{2\pi \mathrm{i}} f_{m,0}^{[3]}(z), \qquad m = 1, 2, \dots,$$

and integrate around the contour Γ_2. Then, by (2.3.83), we obtain

$$C_{1,0}^{[0]} = \int_{\Gamma_2} f(z) f_{1,0}^{[0]}(z)\,\mathrm{d}z, \qquad C_{m,0}^{[1]} = \int_{\Gamma_2} f(z) f_{m,0}^{[1]}(z)\,\mathrm{d}z,$$

$$C_{m,0}^{[3]} = \int_{\Gamma_2} f(z) f_{m,0}^{[3]}(z)\,\mathrm{d}z. \tag{2.3.84}$$

Taking into account that $\psi_3(z) = 0$, $\varphi_3(z) = 0$, $f(z) = 0$ in $\overline{P}$, we obtain from (2.3.82) and (2.3.84)

$$a_{1,0}^{[0]} = b_{1,0}^{[0]} = C_{1,0}^{[0]} = 0,$$

$$a_{m,0}^{[3]} = b_{m,0}^{[3]} = C_{m,0}^{[3]} = C_{m,0}^{[1]} = 0, \qquad m = 1, 2, \dots.$$

Hence it follows that $\alpha_m = \beta_m = c_m = d_m = 0$ for all $m = 0, 1, 2, \dots$. Returning to formula (2.3.71) we see that $u(x, y) = 0$ in D_1. Theorem 2.3.6 is proved. □

Suppose now that $f(x)$, $\psi_j(y)$, and $\varphi_j(x)$, $j = 1, 2, 3$, are known with accuracy δ; i.e., we know the functions $f_\delta(x)$, $\psi_{j\delta}(y)$, and $\varphi_{j\delta}(x)$ such that

$$\|f(x) - f_\delta(x)\|_{L_2} \le \delta,$$

$$\|\psi_j(y) - \psi_{j\delta}(y)\|_{L_2} \le \delta, \qquad \|\varphi_j(x) - \varphi_{j\delta}(x)\|_{L_2} \le \delta.$$

As an approximate solution of the problem (2.3.65)–(2.3.69) we take the function

$$u_\delta^\alpha(x, y) = \sum_{m=0}^{\infty} \Big\{ \big[\alpha_m C_m(y) + \beta_m D_m(y) + \bar{\alpha}_m \bar{C}_m(y) + \bar{\beta}_m \bar{D}_m(y)\big] \cos \frac{m\pi x}{b}$$

$$+ \left[c_m C_m(x) + d_m D_m(x) + \bar{c}_m \bar{C}_m(x) + \bar{d}_m \bar{D}_m(x)\right] \cos \frac{m\pi y}{b} \Big\}$$

where

$$C_0(t) = 1 - \frac{3}{b^2} t^2 + \frac{2}{b^3} t^3, \qquad D_0(t) = \frac{3}{b^2} t^2 - \frac{2}{b^3} t^3,$$

$$\bar{C}_0(t) = t - \frac{3}{b} t^2 + \frac{2}{b^2} t^3, \qquad \bar{D}_0(t) = \frac{3}{b^2} t^3 - \frac{2}{b} t^2;$$

the functions $C_m(t)$, $D_m(t)$ are defined by formulas (2.3.73), (2.3.74);

$$\bar{C}_m(t) = \left[m\pi(b-t) \sinh \frac{m\pi t}{b} - t \sinh m\pi \cdot \frac{m\pi}{b}(b-t)\right] \frac{1}{\sinh^2 m\pi - m^2\pi^2},$$

$$\bar{D}_m(t) = \left[(b-t) \sinh m\pi \cdot \sinh \frac{m\pi t}{b} - m\pi t \sinh \frac{m\pi(b-t)}{b}\right] \times \frac{1}{\sinh^2 m\pi - m^2\pi^2};$$

α is the regularization parameter; α_m, β_m, c_m, d_m, $\bar{\alpha}_m$, $\bar{\beta}_m$, $\bar{c}_m$, and $\bar{d}_m$ are the Fourier coefficients of the functions $T^{\alpha}_{1\delta}(x)$, $T^{\alpha}_{2\delta}(x)$, $T^{\alpha}_{3\delta}(x)$, $\varphi_{1\delta}(x)$, $\varphi_{2\delta}(x)$, $\psi_{1\delta}(x)$, and $\psi_{2\delta}(x)$, respectively. The functions $T^{\alpha}_{k\delta}(x)$, $k = 1, 2, 3$, are an approximate solution of the system of integral equations of first kind

$$\sum_{k=1}^{3} \int_0^b R_{jk}(x,s) T_k(s)\, ds = F_{j\delta}(x), \quad j = 1, 2, 3, \quad a_1 \le x \le a_2, \tag{2.3.85}$$

where

$$R_{11}(x,s) = \frac{2}{b} \sum_{m=0}^{\infty} C_m(a_1) \cos \frac{m\pi x}{b} \cos \frac{m\pi s}{b},$$

$$R_{12}(x,s) = \frac{2}{b} \sum_{m=0}^{\infty} \left[C_m(x) \cos \frac{m\pi a_1}{b} \cos \frac{m\pi s}{b} + D_m(a_1) \cos \frac{m\pi x}{b} \cos \frac{m\pi s}{b}\right],$$

$$R_{13}(x,s) = \frac{2}{b} \sum_{m=0}^{\infty} D_m(x) \cos \frac{m\pi a_1}{b} \cos \frac{m\pi s}{b},$$

$$R_{21}(x,s) = \frac{2}{b} \sum_{m=0}^{\infty} C_m(x) \cos \frac{m\pi a_2}{b} \cos \frac{m\pi s}{b},$$

$$R_{22}(x,s) = \frac{2}{b} \sum_{m=0}^{\infty} \left[C_m(a_2) \cos \frac{m\pi x}{b} \cos \frac{m\pi s}{b} + D_m(x) \cos \frac{m\pi a_2}{b} \cos \frac{m\pi s}{b}\right],$$

$$R_{23}(x,s) = \frac{2}{b}\sum_{m=0}^{\infty} D_m(a_2)\cos\frac{m\pi x}{b}\cos\frac{m\pi s}{b},$$

$$R_{31}(x,s) = \frac{2}{b}\sum_{m=0}^{\infty} C_m(x)\cos\frac{m\pi x}{b}\cos\frac{m\pi s}{b},$$

$$R_{32}(x,s) = \frac{2}{b}\sum_{m=0}^{\infty} [C_m(x)+D_m(x)]\cos\frac{m\pi x}{b}\cos\frac{m\pi s}{b},$$

$$R_{33}(x,s) = \frac{2}{b}\sum_{m=0}^{\infty} D_m(x)\cos\frac{m\pi x}{b}\cos\frac{m\pi s}{b},$$

$$\begin{aligned} F_{1\delta}(x) = \varphi_{1\delta}(x) - \sum_{m=0}^{\infty}\Big\{&\big[\bar{\alpha}_m\bar{C}_m(a_1)+\bar{\beta}_m D_m(a_1)\big]\cos\frac{m\pi x}{b} \\ &+ \big[\bar{c}_m\bar{C}_m(x)+\bar{d}_m\bar{D}_m(x)\big]\cos\frac{m\pi a_1}{b}\Big\}, \\ F_{2\delta}(x) = \psi_{3\delta}(x) - \sum_{m=0}^{\infty}\Big\{&\big[\bar{\alpha}_m\bar{C}_m(x)+\bar{\beta}_m D_m(a_1)\big]\cos\frac{m\pi a_2}{b} \\ &+ \big[\bar{c}_m\bar{C}_m(a_2)+\bar{d}_m\bar{D}_m(x)\big]\cos\frac{m\pi x}{b}\Big\}, \\ F_{3\delta}(x) = f_\delta(x) - \sum_{m=0}^{\infty}\Big\{&[\bar{\alpha}_m+\bar{c}_m]\bar{C}_m(x)+[\bar{\beta}_m+\bar{d}_m]\bar{D}_m(x)\Big\}\cos\frac{m\pi x}{b}. \end{aligned}$$

Now we introduce the functions

$$F_\delta(x) = \begin{cases} F_{1\delta}(x), & a_1 \le x \le a_2, \\ F_{2\delta}(x+a_1-a_2), & a_2 < x \le 2a_2 - a_1, \\ F_{3\delta}(x+2a_1-2a_2), & 2a_2 - a_1 < x \le 3a_2 - 2a_1, \end{cases}$$

$$T(s) = \begin{cases} T_1(s), & 0 \le s \le b, \\ T_2(s-b), & b < s \le 2b, \\ T_3(s-2b), & 2b < s \le 3b, \end{cases}$$

$$R(x,s) = \begin{cases} R_{11}(x,s), & a_1 \le x \le a_2,\ 0 \le s \le b, \\ R_{1,j+1}(x, s - jb), & a_1 \le x \le a_2, \\ & jb < s \le (j+1)b,\ j = 1,2, \\ R_{21}(x + a_1 - a_2, s), & a_2 < x \le 2a_2 - a_1,\ 0 \le s \le b, \\ R_{2,j+1}(x + a_1 - a_2, s - jb), & a_2 < x \le 2a_2 - a_1, \\ & jb < s \le (j+1)b,\ j = 1,2, \\ R_{31}(x + 2a_1 - 2a_2, s), & 2a_2 - a_1 < x \le 3a_2 - 2a_1, \\ & 0 \le s \le b, \\ R_{3,j+1}(x + 2a_1 - 2a_2, s - jb), & 2a_2 - a_1 < x \le 3a_2 - 2a_1, \\ & jb < s \le (j+1)b,\ j = 1,2. \end{cases}$$

Now we can rewrite system (2.3.85) as one equation

$$\int_0^{3b} R(x,s)T(s)\,\mathrm{d}s = F_\delta(x), \qquad a_1 \le x \le 3a_2 - 2a_1. \tag{2.3.86}$$

An approximate solution of this equation is constructed by the Tikhonov regularization method (Tikhonov and Arsenin, 1977).

The following estimate holds

$$\|u(x,y) - u_\delta^\alpha(x,y)\|_{C(D_1)} \le \sum_{k=1}^{3} h_k \bar{\omega}_k(\delta) + h_4\delta,$$

where

$$\bar{\omega}_k(\delta) \ge \|T_k(s) - T_{k\delta}^\alpha(s)\|_{C(0,b)}, \qquad k = 1,2,3,$$

moreover, $\bar{\omega}_k(\delta) \to 0$ for $\delta \to 0$;

$$h_k = \max_{(x,y)\in D_1} \left\{ \int_0^b |R_k(x,y,s)|\,\mathrm{d}s \right\}, \qquad k = 1,2,3,$$

$$h_4 = \sum_{k=4}^{7} \left\{ \max_{(x,y)\in D_1} \left[\int_0^b R_k^2(x,y,s)\,\mathrm{d}s \right]^{1/2} \right\}.$$

Here

$$R_1(x,y,s) = \frac{2}{b} \sum_{m=0}^{\infty} C_m(y) \cos\frac{m\pi x}{b} \cos\frac{m\pi s}{b},$$

$$R_2(x,y,s) = \frac{2}{b} \sum_{m=0}^{\infty} \left[C_m(x) \cos\frac{m\pi y}{b} + D_m(y) \cos\frac{m\pi x}{b} \right] \cos\frac{m\pi s}{b},$$

$$R_3(x,y,s) = \frac{2}{b}\sum_{m=0}^{\infty} D_m(x)\cos\frac{m\pi y}{b}\cos\frac{m\pi s}{b},$$

$$R_4(x,y,s) = \frac{2}{b}\sum_{m=0}^{\infty} \bar{C}_m(y)\cos\frac{m\pi x}{b}\cos\frac{m\pi s}{b},$$

$$R_5(x,y,s) = \frac{2}{b}\sum_{m=0}^{\infty} \bar{D}_m(y)\cos\frac{m\pi x}{b}\cos\frac{m\pi s}{b},$$

$$R_6(x,y,s) = R_4(y,x,s), \qquad R_7(x,y,s) = R_5(y,x,s).$$

Case C. Let $D_1 = \{(x,y) \mid 0 < x < a,\ 0 < y < b\}$, $S_1 = \partial D_1$, and let ν be the external normal to S_1.

Problem 2.3.3. Find the function $u(x,y)$ in D_1 satisfying the conditions

$$\Delta^2 u(x,y) = 0 \quad \text{in} \quad D_1, \tag{2.3.87}$$

$$\frac{\partial u(0,y)}{\partial x} = \frac{\partial u(a,y)}{\partial x} = 0, \qquad 0 \le y \le b, \tag{2.3.88}$$

$$\frac{\partial}{\partial x}\Delta u(0,y) = \frac{\partial}{\partial x}\Delta u(a,y) = 0, \qquad 0 \le y \le b, \tag{2.3.89}$$

$$\frac{\partial u(x,0)}{\partial y} = \varphi_1(x), \qquad \frac{\partial u(x,b)}{\partial y} = \varphi_2(x), \qquad 0 \le x \le a, \tag{2.3.90}$$

$$u(x,b_1) = \varphi_3(x), \qquad u(x,b_2) = \varphi_4(x), \qquad 0 \le x \le a, \tag{2.3.91}$$

where $0 < b_1 < b_2 < b$; $\varphi_j(x)$, $j = \overline{1,4}$, are given functions.

Let us show that in this problem the solution does not depend continuously on the data. Consider the function

$$u_m(x,y) = \varepsilon\, \frac{\widetilde{E}_m(b_2)\widetilde{F}_m(y) - \widetilde{E}_m(y)\widetilde{F}_m(b_2)}{\widetilde{E}_m(b_2)\widetilde{F}_m(b_1) - \widetilde{E}_m(b_1)\widetilde{F}_m(b_2)}\cos mx,$$

where

$$\begin{aligned}\widetilde{E}_m(y) &= \sinh m\pi\big[\sinh m(\pi - y) + my\cosh m(\pi - y)\big]\\ &\quad - m\pi\big[\sinh my + m(\pi - y)\cosh my\big],\\ \widetilde{F}_m(y) &= \sinh m\pi\,\sinh my + m(\pi\cosh m\pi\,\sinh my - y\sinh m\pi\,\cosh my)\\ &\quad - m^2\pi y\cosh m(\pi - y).\end{aligned}$$

The functions $u_m(x, y)$ solve the problem (2.3.87)–(2.3.91) for

$$\varphi_1(x) = \varphi_2(x) = \varphi_4(x) = 0, \qquad \varphi_3(x) = \varepsilon \cos mx, \qquad a = b = \pi.$$

Evidently, for each pair of spaces and for any constants $\bar{\varepsilon} > 0$ and $c > 0$ we may choose ε and m such that

$$\|\varepsilon \cos mx\| < \bar{\varepsilon}, \qquad \|u_m(x, y)\| > c.$$

The following statement holds.

Theorem 2.3.7. *Let the following conditions hold:*

1) *$u(x, y)$ is biharmonic in D_1;*

2) *$u(x, y)$ and $\frac{\partial}{\partial y}u(x, y)$ for $y = 0, b$ satisfy the Hölder condition relative to x, $0 \le x \le a$, with the index $\theta > 1/2$;*

3) *$\frac{\partial}{\partial x}u(x, y)$ and $\frac{\partial}{\partial x}\Delta u(x, y)$ are equal to zero for $x = 0, a$, $0 \le y \le b$.*

Then the equalities

$$\frac{\partial u(x,0)}{\partial y} = \frac{\partial u(x,b)}{\partial y} = 0, \qquad 0 \le x \le a, \tag{2.3.92}$$

$$u(x, b_1) = u(x, b_2) = 0, \qquad 0 \le x \le a, \tag{2.3.93}$$

imply $u(x, y) = 0$ in D_1.

Proof. Consider the function

$$u(x, y) = \sum_{m=0}^{\infty} \left[\alpha_m E_m(y) + \beta_m F_m(y) + c_m \bar{E}_m(y) + d_m \bar{F}_m(y)\right] \cos \frac{m\pi x}{a}, \tag{2.3.94}$$

where

$$E_0(y) = 1 - \frac{3}{b^2}y^2 + \frac{2}{b^3}y^3, \qquad F_0(y) = \frac{3}{b^2}y^2 - \frac{2}{b^3}y^3,$$

$$\bar{E}_0(y) = 1 - \frac{2}{b}y^2 + \frac{2}{b^2}y^3, \qquad \bar{F}_0(y) = \frac{3}{b^2}y^3 - \frac{1}{b}y^2,$$

$$E_m(y) = \left[\sinh \frac{m\pi b}{a}\left(\sinh \frac{m\pi(b-y)}{a} + \frac{m\pi y}{a} \cosh \frac{m\pi(b-y)}{a}\right)\right.$$

$$- \frac{m\pi b}{a}\Big(\sinh\frac{m\pi y}{a} + \frac{m\pi}{a}(b-y)\cosh\frac{m\pi y}{a}\Big)\Big]\Theta,$$

$$F_m(y) = \Big[\sinh\frac{m\pi b}{a}\sinh\frac{m\pi y}{a} - \frac{m^2\pi^2 by}{a^2}\cosh\frac{m\pi(b-y)}{a}$$

$$+ \frac{m\pi}{a}\Big(b\cosh\frac{m\pi b}{a}\sinh\frac{m\pi y}{a} - y\sinh\frac{m\pi b}{a}\cosh\frac{m\pi y}{a}\Big)\Big]\Theta,$$

$$\bar{E}_m(y) = -\Big[\frac{m\pi b}{a}(b-y)\sinh\frac{m\pi y}{a} - y\sinh\frac{m\pi b}{a}\sinh\frac{m\pi(b-y)}{a}\Big]\Theta,$$

$$\bar{F}_m(y) = -\Big[(b-y)\sinh\frac{m\pi b}{a}\sinh\frac{m\pi y}{a} - \frac{m\pi by}{a}\sinh\frac{m\pi(b-y)}{a}\Big]\Theta,$$

$$\Theta = \Big[\sinh^2\frac{m\pi b}{a} - \frac{m^2\pi^2 b^2}{a^2}\Big]^{-1},$$

$$\alpha_m = \frac{2}{a}\int_0^a u(x,0)\cos\frac{m\pi x}{a}\,dx, \qquad \beta_m = \frac{2}{a}\int_0^a u(x,b)\cos\frac{m\pi x}{a}\,dx,$$

$$c_m = \frac{2}{a}\int_0^a \frac{\partial u(x,0)}{\partial y}\cos\frac{m\pi x}{a}\,dx, \qquad d_m = \frac{2}{a}\int_0^a \frac{\partial u(x,b)}{\partial y}\cos\frac{m\pi x}{a}\,dx.$$

Since $u(x,0)$, $u(x,b)$, $\partial u(x,0)/\partial y$, and $\partial u(x,b)/\partial y$ satisfy the Hölder condition with the index $\theta > 1/2$, therefore, by the Bernstein theorem (Zigmund, 1959), the series of the Fourier coefficients of these functions converge absolutely.

Hence and from (2.3.94) it follows that the function $u(x,y)$ is biharmonic in D_1 and satisfies condition *3)*. The functions $u(x,y)$ and $\partial u(x,y)/\partial y$ are continuous for $y = 0, b$, $0 \le x \le a$ and

$$\lim_{y\to 0} u(x,y) = u(x,0), \qquad \lim_{y\to b} u(x,y) = u(x,b),$$

$$\lim_{y\to 0}\frac{\partial u(x,y)}{\partial y} = \frac{\partial u(x,0)}{\partial y}, \qquad \lim_{y\to b}\frac{\partial u(x,y)}{\partial y} = \frac{\partial u(x,b)}{\partial y}.$$

Now, from (2.3.92) and (2.3.93), we obtain

$$\alpha_m E_m(b_j) + \beta_m F_m(b_j) = 0, \qquad j = 1,2, \quad m = 0,1,\dots.$$

Consequently,

$$\alpha_m = \frac{2}{a}\int_0^a u(x,0)\cos\frac{m\pi x}{a}\,dx = 0, \quad \beta_m = \frac{2}{a}\int_0^a u(x,b)\cos\frac{m\pi x}{a}\,dx = 0,$$

$$m = 0,1,2,\dots.$$

Therefore,

$$u(x,0) = u(x,b) = 0, \qquad 0 \le x \le a. \tag{2.3.95}$$

Let

$$D_{1\varepsilon} = \{(x,y) \mid \varepsilon < x < a-\varepsilon,\ \varepsilon < y < b-\varepsilon\}, \quad \varepsilon > 0, \quad S_{1\varepsilon} = \partial D_{1\varepsilon}.$$

Applying the Green formula

$$\iint_{D_{1\varepsilon}} (\psi\Delta\varphi - \varphi\Delta\psi)\, dx\, dy = \int_{S_1} \Big(\psi\frac{\partial\varphi}{\partial\nu} - \varphi\frac{\partial\psi}{\partial\nu}\Big)\, dS_{1\varepsilon}$$

to the functions $\varphi = u$ and $\psi = \Delta u$, we obtain

$$\iint_{D_{1\varepsilon}} (\Delta u)^2\, dx\, dy = \int_{S_1} \Big(\Delta u\,\frac{\partial u}{\partial\nu} - u\,\frac{\partial\Delta u}{\partial\nu}\Big)\, dS_{1\varepsilon}. \tag{2.3.96}$$

Passing to the limit as $\varepsilon \to 0$ in (2.3.96) and taking into account condition *3)*, (2.3.92), and (2.3.95), we get

$$\iint_{D_1} (\Delta u)^2\, dx\, dy = 0.$$

Hence, we obtain

$$\Delta u(x,y) = 0 \qquad \text{in} \quad D_1.$$

Taking into account that

$$u(x,0) = u(x,b) = 0, \qquad 0 \le x \le a,$$
$$\frac{\partial u(0,y)}{\partial x} = \frac{\partial u(a,y)}{\partial x} = 0, \qquad 0 \le y \le b,$$

we see that $u(x,y) = 0$ in D_1. Theorem 2.3.7 is proved. □

The solution to the problem (2.3.87)–(2.3.91) can be represented as

$$u(x,y) = \sum_{k=1}^{4} u_k(x,y),$$

where $u_k(x,y)$ are the solutions of the problems

$$\Delta^2 u_k(x,y) = 0 \qquad \text{in} \quad D_1, \tag{2.3.97}$$

$$\frac{\partial u_k(0,y)}{\partial x} = \frac{\partial u_k(a,y)}{\partial x} = 0, \qquad 0 \le y \le b, \tag{2.3.98}$$

$$\frac{\partial}{\partial x}\Delta u_k(0,y) = \frac{\partial}{\partial x}\Delta u_k(a,y) = 0, \qquad 0 \le y \le b, \tag{2.3.99}$$

$$\frac{\partial u_k(x,0)}{\partial y} = \varphi_1(x), \qquad \frac{\partial u_k(x,b)}{\partial y} = \varphi_2(x), \qquad 0 \le x \le a, \tag{2.3.100}$$

$$u_k(x,b_1) = \varphi_3(x), \qquad u_k(x,b_2) = \varphi_4(x), \qquad 0 \le x \le a, \tag{2.3.101}$$

where $\varphi_j(x) = 0$ for

$$j \ne k, \qquad j,k = \overline{1,4}. \tag{2.3.102}$$

Consider the problem

$$\Delta^2 u_3(x,y) = 0 \quad \text{in} \quad D_1, \tag{2.3.103}$$

$$\frac{\partial u_3(0,y)}{\partial x} = \frac{\partial u_3(a,y)}{\partial x} = 0, \qquad 0 \le y \le b, \tag{2.3.104}$$

$$\frac{\partial}{\partial x}\Delta u_3(0,y) = \frac{\partial}{\partial x}\Delta u_3(a,y) = 0, \qquad 0 \le y \le b, \tag{2.3.105}$$

$$\frac{\partial u_3(x,0)}{\partial y} = 0, \qquad \frac{\partial u_3(x,b)}{\partial y} = 0, \qquad 0 \le x \le a, \tag{2.3.106}$$

$$u_3(x,b_1) = \varphi_3(x), \qquad u_3(x,b_2) = 0, \qquad 0 \le x \le a. \tag{2.3.107}$$

Let us investigate this problem for conditional well-posedness.

The solution of the problem (2.3.103)–(2.3.107) can be written as

$$u_3(x,y) = \sum_{m=0}^{\infty} f_m E_m^{[3]}(y) \cos\frac{m\pi x}{a}, \tag{2.3.108}$$

where f_m are the Fourier coefficients of the function $\varphi_3(x)$;

$$E_0^{[3]}(y) = \left[b_2^2(3b - 2b_2) - 3by^2 + 2y^3\right]\left[3b(b_2^2 - b_1^2) - 2(b_2^3 - b_1^3)\right]^{-1},$$

$$E_m^{[3]}(y)$$

$$= \Big\{\Big[\frac{m\pi b_2 b}{a}\cosh\frac{m\pi(b_2 - b)}{a} - \Big(\sinh\frac{m\pi b_2}{a} - \frac{m\pi b_2}{a}\cosh\frac{m\pi b_2}{a}\Big)\sinh\frac{m\pi b}{a}$$

$$- \frac{m\pi b}{a}\sinh\frac{m\pi b_2}{a}\cosh\frac{m\pi b}{a}\Big]\cosh\frac{m\pi y}{a}$$

$$+ \Big[\frac{m\pi b}{a}\cosh\frac{m\pi b_2}{a}\cosh\frac{m\pi b}{a} + \Big(\cosh\frac{m\pi b_2}{a} - b_2\sinh\frac{m\pi b_2}{a}\Big)\sinh\frac{m\pi b}{a}\Big]$$

$$\times \left(\sinh\frac{m\pi y}{a} - \frac{m\pi y}{a}\cosh\frac{m\pi y}{a}\right)$$

$$+ \frac{m\pi y}{a}\left(\sinh\frac{m\pi b_2}{a} + \frac{m\pi(b-b_2)}{a}\cosh\frac{m\pi b_2}{a}\right)\sinh\frac{m\pi b}{a}\sinh\frac{m\pi y}{a}\Bigg\}$$

$$\times \Bigg\{\frac{m^2\pi^2}{a^2}\Bigg[(b_2-b)b\cosh\frac{m\pi b_1}{a}\cosh\frac{m\pi b_2}{a}\cosh\frac{m\pi b}{a}$$

$$+ \left(b_1b_2\sinh\frac{m\pi(b_2-b_1)}{a} - (b_2-b_1)b\sinh\frac{m\pi(b_2-b_1)}{a}\right)\sinh\frac{m\pi b}{a}\Bigg]$$

$$- \frac{m\pi b}{a}\sinh\frac{m\pi(b_2-b_1)}{a}\cosh\frac{m\pi b}{a}$$

$$- \left(\sinh\frac{m\pi(b_2-b_1)}{a} - \frac{m\pi(b_1+b_2)}{a}\cosh\frac{m\pi(b_2-b_1)}{a}\right)\sinh\frac{m\pi b}{a}\Bigg\}^{-1}. \tag{2.3.109}$$

The theorem below characterizes the stability of solution of the problem (2.3.103)–(2.3.107).

Theorem 2.3.8. *If the function $u_3(x,y)$ satisfies the relations*

$$\|u_3(x,b)\|_{L_2(0,a)} \le \frac{M}{4}, \tag{2.3.110}$$

$$\|u_3(x,b_1)\|_{L_2(0,a)} \le \varepsilon_3,$$

$$\frac{\partial u_3(0,y)}{\partial x} = \frac{\partial u_3(a,y)}{\partial x} = 0, \qquad 0 \le y \le b,$$

$$\frac{\partial}{\partial x}\Delta u_3(0,y) = \frac{\partial}{\partial x}\Delta u_3(a,y) = 0, \qquad 0 \le y \le b,$$

$$\frac{\partial u_3(x,0)}{\partial y} = \frac{\partial u_3(x,b)}{\partial y} = u_3(x,b_2) = 0, \qquad 0 \le x \le a,$$

then the inequality

$$\|u_3(x,b)\|_{L_2(0,a)} \le \frac{M}{4}\left|\frac{E^{[3]}_{\bar\nu(\varepsilon_3)}(y)}{E^{[3]}_{\bar\nu(\varepsilon_3)}(b)}\right|$$

holds. Here $E^{[3]}_{\bar\nu}(y)$ is defined by formula (2.3.109), and $\bar\nu(\varepsilon_3)$ is the positive root of the equation

$$4|E^{[3]}_{\bar\nu}(b)| = M\varepsilon_3^{-1}. \tag{2.3.111}$$

Note that, by (2.3.109) and (2.3.111), $\bar{\nu}(\varepsilon_3) \to \infty$ for $\varepsilon_3 \to 0$.

Theorem 2.3.8 is proved analogously to Theorem 2.3.2.

We suppose now that the function $\varphi_3(x)$ is known to within δ; i. e., we know the function $\varphi_{3\delta}(x)$ such that

$$\|\varphi_3(x) - \varphi_{3\delta}(x)\|_{L_2(0,a)} \leq \delta.$$

We assume that the problem (2.3.103)–(2.3.107) is set coditionally well-posed, and the set of well-posedness is given by inequality (2.3.110). As an approximate solution of the problem (2.3.103)–(2.3.107) we take the function

$$u_{3n\delta}(x,y) = \sum_{m=0}^{n} \bar{f}_m E_m^{[3]}(y) \cos\frac{m\pi x}{a},$$

where $\bar{f}_m$ are the Fourier coefficients of the function $\varphi_{3\delta}(x)$.

Let us estimate $u_3(x,y) - u_{3n\delta}(x,y)$ in the metric of the space $L_2(0,a)$. Similarly to (2.1.37) we obtain

$$\|u_3(x,y) - u_{3n\delta}(x,y)\|_{L_2(0,a)} \leq |E_n^{[3]}(y)|\delta + \frac{M}{4}\left|\frac{E_{n+1}^{[3]}(y)}{E_{n+1}^{[3]}(b)}\right|.$$

Thus, the problem (2.3.103)–(2.3.107) is investigated in $L_2(0,a)$. Analogously we investigate all other problems (2.3.97)–(2.3.102).

Now, for the problem (2.3.87)–(2.3.91), we can formulate the following results.

Theorem 2.3.9. *If the function $u(x,y)$ satisfies the relations*

$$\|u(x,b)\|_{L_2(0,a)} \leq M,$$

$$\left\|\frac{\partial u(x,0)}{\partial y}\right\|_{L_2(0,a)} \leq \varepsilon_1, \qquad \left\|\frac{\partial u(x,b)}{\partial y}\right\|_{L_2(0,a)} \leq \varepsilon_2,$$

$$\|u(x,b_1)\|_{L_2(0,a)} \leq \varepsilon_3, \qquad \|u(x,b_2)\|_{L_2(0,a)} \leq \varepsilon_4,$$

$$\frac{\partial u(0,y)}{\partial x} = \frac{\partial u(a,y)}{\partial x} = 0, \qquad 0 \leq y \leq b,$$

$$\frac{\partial \Delta u(0,y)}{\partial x} = \frac{\partial \Delta u(a,y)}{\partial x} = 0, \qquad 0 \leq y \leq b,$$

then the inequality

$$\|u(x,y)\|_{L_2(0,a)} \leq \frac{M}{4}\sum_{j=1}^{4}\left|\frac{E^{[j]}_{\bar{\nu}(\varepsilon_j)}(y)}{E^{[3]}_{\bar{\nu}(\varepsilon_j)}(b)}\right| \tag{2.3.112}$$

holds, where $E^{[j]}_{\bar{\nu}}(y)$ are known functions and

$$\left|\frac{E^{[j]}_{\bar{\nu}}(y)}{E^{[j]}_{\bar{\nu}}(b)}\right| \to 0$$

for $\bar{\nu} \to \infty$; $\bar{\nu}(\varepsilon_j)$ are positive roots of the equations

$$4|E^{[j]}_{\bar{\nu}}(b)| = \frac{1}{\varepsilon_j}, \qquad j = \overline{1,4}.$$

Assume that $\varphi_j(x)$, $j = \overline{1,4}$, are known with accuracy δ. This means that we know the functions $\varphi_{j\delta}(x)$ such that

$$\|\varphi_j(x) - \varphi_{j\delta}(x)\|_{L_2(0,a)} \leq \delta.$$

As an approximate solution of the problem (2.3.87)–(2.3.91) we take the function

$$u_{n\delta}(x,y) = \sum_{k=1}^{4} u_{kn\delta}(x,y),$$

where $u_{kn\delta}(x,y)$ are approximate solutions of the problem (2.3.97)–(2.3.102).

The following estimate holds

$$\|u_{n\delta}(x,y) - u(x,y)\|_{L_2(0,a)} \leq \tilde{\omega}(n,\delta,y), \tag{2.3.113}$$

where

$$\tilde{\omega}(n,\delta,y) = \sum_{k=1}^{4}\left[|E^{[k]}_{n}(y)|\delta + \frac{M}{4}\left|\frac{E^{[k]}_{n+1}(y)}{E^{[k]}_{n+1}(b)}\right|\right].$$

For fixed accuracy δ of the approximate data, the value of the parameter n for which $\inf\tilde{\omega}(n,\delta,y)$ is achieved will be optimal in the sense of the estimate (2.3.113).

Suppose now that the constant M in inequality (2.3.112) which determines the set of well-posedness of the problem (2.3.87)–(2.3.91) is unknown.

Then as an approximate solution of the problem (2.3.87)–(2.3.91) we take the function

$$u_\delta^\alpha(x,y) = \sum_{m=0}^{\infty} \left[\bar{\alpha}_m E_m(y) + \bar{\beta}_m F_m(y) + \bar{c}_m \bar{E}_m(y) + \bar{d}_m \bar{F}_m(y)\right] \cos\frac{m\pi x}{a},$$

where α is the regularization parameter; $\bar{\alpha}_m$, $\bar{\beta}_m$, $\bar{c}_m$, and $\bar{d}_m$ are the Fourier coefficients of the functions $q_{1\delta}^\alpha(x)$, $q_{2\delta}^\alpha(x)$, $\varphi_{1\delta}^\alpha(x)$, and $\varphi_{2\delta}^\alpha(x)$; and $q_{1\delta}^\alpha(x)$, $q_{2\delta}^\alpha(x)$ are an approximate solution of the system of integral equations of first kind

$$\sum_{k=1}^{2} \int_0^a Q_{jk}(x,s) q_k(s)\, \mathrm{d}s = f_{j\delta}(x), \quad j = 1,2, \quad 0 \le x \le a. \tag{2.3.114}$$

Here

$$Q_{j1}(x,s) = \frac{2}{a} \sum_{m=0}^{\infty} E_m(b_j) \cos\frac{m\pi x}{a} \cos\frac{m\pi s}{a},$$

$$Q_{j2}(x,s) = \frac{2}{a} \sum_{m=0}^{\infty} F_m(b_j) \cos\frac{m\pi x}{a} \cos\frac{m\pi s}{a},$$

$$f_{j\delta}(x) = \varphi_{2+j,\delta}(x) - \sum_{m=0}^{\infty} \left[\bar{c}_m \bar{E}_m(b_j) + \bar{d}_m \bar{F}_m(b_j)\right] \cos\frac{m\pi x}{a}, \qquad j = 1,2.$$

Setting

$$f_\delta(x) = \begin{cases} f_{1\delta}(x), & 0 \le x \le a, \\ f_{2\delta}(x-a), & a < x \le 2a, \end{cases}$$

$$q(s) = \begin{cases} q_1(s), & 0 \le s \le a, \\ q_2(s-a), & a < s \le 2a, \end{cases}$$

$$Q(x,s) = \begin{cases} Q_{11}(x,s), & 0 \le x \le a, \quad 0 \le s \le a, \\ Q_{12}(x,s-a), & 0 \le x \le a, \quad a < s \le 2a, \\ Q_{21}(x-a,s), & a < x \le 2a, \quad 0 \le s \le a, \\ Q_{22}(x-a,s-a), & a < x \le 2a, \quad a < s \le 2a, \end{cases}$$

we rewrite system (2.3.114) as one equation

$$\int_0^{2a} Q(x,s) q(s)\, \mathrm{d}s = f_\delta(x), \qquad 0 \le x \le 2a. \tag{2.3.115}$$

An approximate solution of equation (2.3.115) is constructed by the Tikhonov regularization method (Tikhonov and Arsenin, 1977)

$$\|u(x,y) - u_\delta^\alpha(x,y)\|_{L_2(0,a)} \le \sum_{k=1}^{2} \eta_k(y)\tilde{\omega}_k(\delta) + \eta_3(y)\delta,$$

where

$$\tilde{\omega}_k(\delta) \ge \|q_k(s) - q_{k\delta}^\alpha(s)\|_{C(0,a)}, \qquad k = 1,2,$$

$\tilde{\omega}_k(\delta) \to 0$ for $\delta \to 0$;

$$\eta_k(y) = \sqrt{\alpha}\left[\int_0^a \int_0^a Q_k^2(x,y,s)\,\mathrm{d}x\,\mathrm{d}s\right]^{1/2}, \qquad k = 1,2,$$

$$\eta_3(y) = \sum_{k=3}^{4}\left[\int_0^a \int_0^a Q_k^2(x,y,s)\,\mathrm{d}x\,\mathrm{d}s\right]^{1/2},$$

$$Q_1(x,y,s) = \frac{2}{a}\sum_{m=0}^{\infty} E_m(y)\cos\frac{m\pi x}{a}\cos\frac{m\pi s}{a},$$

$$Q_2(x,y,s) = \frac{2}{a}\sum_{m=0}^{\infty} F_m(y)\cos\frac{m\pi x}{a}\cos\frac{m\pi s}{a},$$

$$Q_3(x,y,s) = \frac{2}{a}\sum_{m=0}^{\infty} \bar{E}_m(y)\cos\frac{m\pi x}{a}\cos\frac{m\pi s}{a},$$

$$Q_4(x,y,s) = \frac{2}{a}\sum_{m=0}^{\infty} \bar{F}_m(y)\cos\frac{m\pi x}{a}\cos\frac{m\pi s}{a}.$$

Thus, we have investigated the problem (2.3.87)–(2.3.91) completely.

Chapter 3.

The third internal boundary value problem

We shall seek the solution of IBVP 3 as the sum of three potentials: simple, double, and logarithmic

$$u(p) = \int_{S_1''} \Big(\mu_1(q) \ln|p-q| - \mu_2(q) \frac{\partial}{\partial n_q} \ln|p-q|\Big)\, \mathrm{d}_q S_1'' + \int_{D_1} W(Q) \ln|Q-p|\, \mathrm{d}Q \qquad (3.0.1)$$

where

$$W(Q) = \int_{S_1''} \mu_3(q) \ln|Q-q|\, \mathrm{d}_q S_1'', \qquad S_1'' = S_1 \setminus S_1'.$$

By condition (2) for $j = 2$, condition (2) for $j = 1$ with S_1' instead of S_1, and condition (4), we arrive at the following system of integral Fredholm equations of first kind

$$\int_{S_1''} \Big[\mu_1(q) \ln|p-q| - \mu_2(q) \frac{\partial}{\partial n_q} \ln|p-q| + \mu_3(q) K(p,q)\Big]\, \mathrm{d}_q S_1'' = \Phi_1(p), \qquad p \in S_1',$$

$$\int_{S_1''} \Big[\mu_1(q) \frac{\partial}{\partial n_p} \ln|p-q| - \mu_1(q) \frac{\partial^2}{\partial n_q\, \partial n_p} \ln|p-q| + \mu_3(q) K(p,q)\Big]\, \mathrm{d}_q S_1'' = \Phi_2(p), \qquad p \in S_1',$$

$$\int_{S_1''} \left[\mu_1(q) \ln|p-q| - \mu_2(q)\frac{\partial}{\partial n_q} \ln|p-q| + \mu_3(q)K(p,q)\right] \mathrm{d}_q S_1'' = \Phi_3(p),$$
$$p \in S_2, \tag{3.0.2}$$

where

$$K(p,q) = \int_{D_1} \ln|p-Q| \ln|q-Q| \, \mathrm{d}Q.$$

For IBVP 3, theorems analogous to Theorems 1.0.1, 1.0.2, and 1.2.1 hold, and we can construct an example of non-uniqueness similar to that in Section 1.1.

3.1. IBVP 3 IN A CONVEX DOMAIN

We suppose that $r = R_2 = \text{const}$. Then, using representation (2.1.4), from (2)–(4) we pass to the following problems

$$\Delta u_1(r,\varphi) = 0 \quad \text{in} \quad D_1, \tag{3.1.1}$$
$$u_1(R_2,\varphi) = \Phi_2(\varphi), \quad \varphi \in [0, 2\pi], \tag{3.1.2}$$
$$\Delta u_2(r,\varphi) = 0 \quad \text{in} \quad D_1, \tag{3.1.3}$$
$$u_2(R_1(\varphi),\varphi) = [R_1^2(\varphi) - R_2^2]^{-1}[\Phi_1(\varphi) - u_1(R_1(\varphi),\varphi)], \tag{3.1.4}$$
$$\left.\frac{\partial u_2}{\partial n}\right|_{r=R_1(\varphi)} = i^{-1}(\varphi)\left[\frac{\Phi_3(\varphi) - \left.\dfrac{\partial u_1}{\partial n}\right|_{r=R_1(\varphi)}}{2R_1(\varphi)\left.\dfrac{\partial r}{\partial n}\right|_{r=R_1(\varphi)}} - u_2(R_1(\varphi),\varphi)\right], \tag{3.1.5}$$

where

$$\varphi \in (-\alpha, \alpha), \qquad \alpha < \pi, \qquad i(\varphi) = \frac{R_1^2(\varphi) - R_2^2}{2R_1(\varphi)\left.\dfrac{\partial r}{\partial n}\right|_{r=R_1(\varphi)}}.$$

The problem (3.1.1), (3.1.2) is the problem of harmonic continuation and the problem (3.1.3)–(3.1.5) is the Cauchy problem for the Laplace equation.

The problem (3.1.1), (3.1.2) was investigated in detail in Chapters 1, 2. The equivalence of the problem (3.1.3)–(3.1.5) and the problem of analytic continuation across S_1' was established in (Lavrent'ev, 1981, p. 42).

Theorem 3.1.1. *If*

$$\|u\|_{L_2(S_1')} + \left\|\frac{\partial u}{\partial n}\right\|_{L_2(S_1')} \le \varepsilon, \qquad \|u\|_{L_2(S_2)} \le \varepsilon,$$

$$\left\| \frac{\partial^2 u_1}{\partial r^2} \right\|_{L_2(S_1')} \leq M, \qquad \|u_2\|_{L_2(S_1'')} + \left\| \frac{\partial u_2}{\partial n} \right\|_{L_2(S_1'')} \leq M,$$

then

$$\|u\|_{L_2(D_1)} \leq \frac{MR_1^2 - (r/R_1)^{\lambda(\varepsilon)}}{\lambda(\varepsilon)[\lambda(\varepsilon)-1]} + M^{1-\omega}\left\{ \frac{\max R_1(\varphi)}{\min R_1^2 - R_2^2}\left[2\varepsilon + \frac{MR_1^2[2r+\lambda(\varepsilon)]}{\lambda(\varepsilon)[\lambda(\varepsilon)-1](r/R_1)^{\lambda(\varepsilon)}}\right]\right\}^{\omega},$$

where

$$\frac{\partial^2 u_1}{\partial r^2} = \frac{\partial u}{\partial r^2}\,\frac{3r^2+R_2^2}{4r^2}\Delta u + \frac{r^2-R_2^2}{4r^2}\int_0^r \Delta u \, \mathrm{d}\tau - \frac{r^2-R_2^2}{4r^2}\frac{\partial}{\partial r}\Delta u,$$

$$u_2 = \frac{1}{4r}\int_0^r \Delta u \, \mathrm{d}\tau,$$

$\lambda(\varepsilon)$ is the root of the equation

$$k(k-1)\left(\frac{R_1}{R_2}\right)^k = \frac{MR_1^2}{\varepsilon},$$

and ω is the harmonic measure of the contour S_1'.

The proof of this theorem is analogous to that of Theorem 2.1.1.

3.2. IBVP 3 IN A RECTANGLE

Let D_1 and D_2 ($D_1 \supset D_2$) be the rectangles

$$D_1 = \{(x,y) \mid 0 < x < x_3,\ 0 < y < y_3\},$$
$$D_2 = \{(x,y) \mid x_1 < x < x_2,\ y_1 < y < y_2\}.$$

IBVP 3. We need to find a function $u(x,y)$ in D_1 satisfying the conditions

$$\Delta^2 u(x,y) = 0 \quad \text{in} \quad D_1, \tag{3.2.1}$$

$$u(x,0) = \varphi_1(x), \qquad u_y'(x,0) = \varphi_2(x), \qquad 0 \leq x \leq x_3, \tag{3.2.2}$$

$$u(0,y) = \psi_1(y), \qquad u_x'(0,y) = \psi_2(y), \qquad 0 \leq y \leq y_3, \tag{3.2.3}$$

$$u(x,y_1) = \varphi_3(x), \qquad u(x,y_2) = \varphi_4(x), \qquad x_1 \leq x \leq x_2, \tag{3.2.4}$$

$$u(x_1,y) = \psi_3(y), \qquad u(x_2,y) = \psi_4(y), \qquad y_1 \leq y \leq y_2, \tag{3.2.5}$$

where $\varphi_j(x)$ and $\psi_j(y)$, $j = \overline{1,4}$, are given functions.

We shall seek the function $u(x,y)$ biharmonic in D_1 in the form

$$
\begin{aligned}
u(x,y) = a_0^{(0)} &+ a_0^{(1)}x + b_0^{(1)}y + a_0^{(2)}(x^2+y^2)\\
&+ \sum_{k=1}^{\infty}\Big\{\sinh^{-1}\frac{k\pi y_3}{x_3}\Big[a_k^{(1)}\sinh\frac{k\pi y}{x_3} + b_k^{(1)}\sinh\frac{k\pi(y_3-y)}{x_3}\\
&\qquad + x\Big(a_k^{(4)}\sinh\frac{k\pi y}{x_3} + b_k^{(3)}\sinh\frac{k\pi(y_3-y)}{x_3}\Big)\Big]\sin\frac{k\pi x}{x_3}\\
&\qquad + \sinh^{-1}\frac{k\pi x_3}{y_3}\Big[c_k^{(1)}\sinh\frac{k\pi x}{y_3} + c_k^{(2)}\sinh\frac{k\pi(x_3-x)}{y_3}\\
&\qquad + x\Big(a_k^{(4)}\sinh\frac{k\pi x}{y_3} + b_k^{(4)}\sinh\frac{k\pi(x_3-x)}{y_3}\Big)\Big]\sin\frac{k\pi y}{x_3}\\
&\qquad + \cosh^{-1}\frac{k\pi y_3}{x_3}\Big[a_k^{(2)}\cosh\frac{k\pi x}{y_3} + b_k^{(2)}\cosh\frac{k\pi(y_3-y)}{y_3}\\
&\qquad + y\Big(a_k^{(3)}\cosh\frac{k\pi y}{x_3} + b_k^{(3)}\cosh\frac{k\pi(y_3-y)}{y_3}\Big)\Big]\cos\frac{k\pi x}{x_3}\\
&\qquad + \cosh^{-1}\frac{k\pi x_3}{y_3}\Big[c_k^{(3)}\cosh\frac{k\pi x}{y_3} + c_k^{(4)}\cosh\frac{k\pi(x_3-x)}{y_3}\\
&\qquad + y\Big(a_k^{(4)}\cosh\frac{k\pi x}{y_3} + b_k^{(4)}\cosh\frac{k\pi(x_3-x)}{y_3}\Big)\Big]\cos\frac{k\pi y}{y_3}\Big\},
\end{aligned}
\tag{3.2.6}
$$

where

$$a_0^{(0)},\ a_0^{(1)},\ a_0^{(2)},\ b_0^{(1)},\ a_k^{(i)},\ b_k^{(i)},\ c_k^{(i)},\quad i=\overline{1,4},\quad k=1,2,\dots,$$

are arbitrary constants such that

$$|a_k^{(i)}|,\ |b_k^{(i)}|,\ |c_k^{(i)}| \le ce^{-kd}; \tag{3.2.7}$$

c and d are non-negative numbers.

The following statement holds.

Theorem 3.2.1. *If*

$$d > \max\Big\{\frac{x_3}{y_3}, \frac{y_3}{x_3}\Big\}\pi, \tag{3.2.8}$$

then the equalitites

$$u(x,0) = u'_y(x,0) = 0, \qquad 0 \le x \le x_3, \tag{3.2.9}$$

$$u(0,y) = u'_x(0,y) = 0, \qquad 0 \le y \le y_3, \tag{3.2.10}$$

$$u(x,y_1) = u(x,y_2) = 0, \qquad x_1 \le x \le x_2, \tag{3.2.11}$$

$$u(x_1,y) = u(x_2,y) = 0, \qquad y_1 \le y \le y_2, \tag{3.2.12}$$

imply that $u(x,y) = 0$ *in* D_1.

Proof. We shall use some statements from the theory of Dirichlet series (Leontiev, 1976) or the Jensen inequality for entire functions.

As follows from (3.2.11) and (3.2.12), the functions

$$\begin{aligned}
\varphi_j(x) &= a_0^{(0)} + a_0^{(1)}x + b_0^{(1)}y_{j-2} + a_0^{(2)}(x^2 + y_{j-2}^2) \\
&+ \sum_{k=1}^{\infty}\Big\{\sinh^{-1}\frac{k\pi y_3}{x_3}\Big[a_k^{(1)}\sinh\frac{k\pi y_{j-2}}{x_3} + b_k^{(1)}\sinh\frac{k\pi(y_3-y_{j-2})}{x_3} \\
&\qquad + x\Big(a_k^{(4)}\sinh\frac{k\pi y_{j-2}}{x_3} + b_k^{(3)}\sinh\frac{k\pi(y_3-y_{j-2})}{x_3}\Big)\Big]\sin\frac{k\pi x}{x_3} \\
&\quad + \sinh^{-1}\frac{k\pi x_3}{y_3}\Big[c_k^{(1)}\sinh\frac{k\pi x}{y_3} + c_k^{(2)}\sinh\frac{k\pi(x_3-x)}{y_3} \\
&\qquad + y_{j-2}\Big(a_k^{(4)}\sinh\frac{k\pi x}{y_3} + b_k^{(4)}\sinh\frac{k\pi(x-x_3)}{y_3}\Big)\Big]\sin\frac{k\pi y_{j-2}}{x_3} \\
&\quad + \cosh^{-1}\frac{k\pi y_3}{x_3}\Big[a_k^{(2)}\cosh\frac{k\pi y_{j-2}}{x_3} + b_k^{(2)}\cosh\frac{k\pi(y_3-y_{j-2})}{x_3} \\
&\qquad + x\Big(a_k^{(3)}\cosh\frac{k\pi y_{j-2}}{x_3} + b_k^{(3)}\cosh\frac{k\pi(y_3-y_{j-2})}{x_2}\Big)\Big]\cos\frac{k\pi x}{x_3} \\
&\quad + \cosh^{-1}\frac{k\pi x_3}{y_3}\Big[c_k^{(3)}\cosh\frac{k\pi x}{y_3} + c_k^{(4)}\cosh\frac{k\pi(x_3-x)}{y_3} \\
&\qquad + y_{j-2}\Big(a_k^{(4)}\cosh\frac{k\pi x}{y_3} + b_k^{(4)}\cosh\frac{k\pi(x_3-x)}{y_3}\Big)\Big]\cos\frac{k\pi y_{j-2}}{y_3}\Big\}, \\
&\qquad j = 3,4,
\end{aligned}$$

vanish on the segment $[x_1, x_2]$, and the functions

$$\psi_j(y) = a_0^{(0)} + a_0^{(1)}x_{j-2} + b_0^{(1)}y + a_0^{(2)}(x_{j-2}^2 + y^2)$$

$$
\begin{aligned}
&+\sum_{k=1}^{\infty}\biggl\{\sinh^{-1}\frac{k\pi y_3}{x_3}\Bigl[a_k^{(1)}\sinh\frac{k\pi y}{x_3}+b_k^{(1)}\sinh\frac{k\pi(y_3-y)}{x_3}\\
&\qquad+x_{j-2}\Bigl(a_k^{(3)}\sinh\frac{k\pi y}{x_3}+b_k^{(3)}\sinh\frac{k\pi(y_3-y)}{x_3}\Bigr)\Bigr]\sin\frac{k\pi x_{j-2}}{x_3}\\
&\quad+\sinh^{-1}\frac{k\pi x_3}{y_3}\Bigl[c_k^{(1)}\sinh\frac{k\pi x_{j-2}}{y_3}+c_k^{(2)}\sinh\frac{k\pi(x_3-x_{j-2})}{y_3}\\
&\qquad+y\Bigl(a_k^{(4)}\sinh\frac{k\pi x_{j-2}}{y_3}+b_k^{(4)}\sinh\frac{k\pi(x_3-x_{j-2})}{y_3}\Bigr)\Bigr]\sin\frac{k\pi y}{x_3}\\
&\quad+\cosh^{-1}\frac{k\pi y_3}{x_3}\Bigl[a_k^{(2)}\cosh\frac{k\pi y}{x_3}+b_k^{(2)}\cosh\frac{k\pi(y_3-y)}{x_3}\\
&\qquad+x_{j-2}\Bigl(a_k^{(3)}\cosh\frac{k\pi y}{x_3}+b_k^{(3)}\cosh\frac{k\pi(y_3-y)}{x_3}\Bigr)\Bigr]\cos\frac{k\pi x_{j-2}}{x_3}\\
&\quad+\cosh^{-1}\frac{k\pi x_3}{y_3}\Bigl[c_k^{(3)}\cosh\frac{k\pi x_{j-2}}{y_3}+c_k^{(4)}\cosh\frac{k\pi(x_3-x_{j-2})}{y_3}\\
&\qquad+y\Bigl(a_k^{(4)}\cosh\frac{k\pi x_{j-2}}{y_3}+b_k^{(4)}\cosh\frac{k\pi(x_3-x_{j-2})}{y_3}\Bigr)\Bigr]\cos\frac{k\pi y}{y_3}\biggr\},\\
&\qquad\qquad j=3,4,
\end{aligned}
$$

vanish on the segment $[y_1, y_2]$.

In virtue of (3.2.7) the functions $\varphi_j(x)$ and $\psi_j(y)$ admit analytical continuation into the rectangles

$$
\overline{P}=\Bigl\{z=x+\mathrm{i}\xi \Bigm| -\frac{y_3}{\pi}d\le \operatorname{Re} z\le \frac{y_3}{\pi}d,\ -\frac{x_3}{\pi}d\le \operatorname{Im} z\le \frac{x_3}{\pi}d\Bigr\},
$$

$$
\overline{\Pi}=\Bigl\{\zeta=y+\mathrm{i}\eta \Bigm| -\frac{x_3}{\pi}d\le \operatorname{Re} \zeta\le \frac{x_3}{\pi}d,\ -\frac{y_3}{\pi}d\le \operatorname{Im} \zeta\le \frac{y_3}{\pi}d\Bigr\},
$$

respectively, by means of the equalities

$$
\begin{aligned}
\varphi_j(z)&=a_{j1,0}^{[0]}+a_{j1,1}^{[0]}z+a_{j1,2}^{[0]}z^2\\
&\quad+\sum_{k=1}^{\infty}\Bigl[a_{jk,0}^{[1]}\mathrm{e}^{k\pi(z/y_3)}+a_{jk,1}^{[1]}z\mathrm{e}^{k\pi(z/y_3)}+a_{jk,0}^{[2]}\mathrm{e}^{-k\pi(z/y_3)}\\
&\qquad+a_{jk,1}^{[2]}z\mathrm{e}^{-k\pi(z/y_3)}+a_{jk,0}^{[3]}\mathrm{e}^{\mathrm{i}k\pi(z/x_3)}+a_{jk,0}^{[4]}\mathrm{e}^{-\mathrm{i}k\pi(z/x_3)}\\
&\qquad+a_{jk,1}^{[3]}z\mathrm{e}^{\mathrm{i}k\pi(z/x_3)}+a_{jk,1}^{[4]}z\mathrm{e}^{-\mathrm{i}k\pi(z/x_3)}\Bigr],
\end{aligned}
$$

$$
a_{j1,0}^{[0]}=a_0^{(0)}+b_0^{(1)}y_{j-2}+a_0^{(2)}y_{j-2}^2,
$$

$$a^{[0]}_{j1,0} = a^{(1)}_0, \qquad a^{[0]}_{j1,2} = a^{(2)}_0,$$

$$\begin{aligned}
a^{[1]}_{jk,0} = \frac{1}{2}\Big[&\sin\frac{k\pi y_{j-2}}{y_3}\sinh^{-1}\frac{k\pi x_3}{y_3}\Big(c^{(1)}_k - c^{(2)}_k \mathrm{e}^{-k\pi(x_3/y_3)}\Big)\\
&+\cos\frac{k\pi y_{j-2}}{y_3}\cosh^{-1}\frac{k\pi x_3}{y_3}\Big(c^{(3)}_k - c^{(4)}_k \mathrm{e}^{-k\pi(x_3/y_3)}\Big)\\
&+y_{j-2}\cos\frac{k\pi y_{j-2}}{y_3}\cosh^{-1}\frac{k\pi x_3}{y_3}\Big(a^{(3)}_k - b^{(4)}_k \mathrm{e}^{-k\pi(x_3/y_3)}\Big)\Big],
\end{aligned}$$

$$a^{[1]}_{jk,1} = \frac{1}{2}\sin\frac{k\pi y_{j-2}}{y_3}\sinh^{-1}\frac{k\pi x_3}{y_3}\Big(a^{(4)}_k - b^{(4)}_k \mathrm{e}^{-k\pi(x_3/y_3)}\Big),$$

$$\begin{aligned}
a^{[2]}_{jk,0} = \frac{1}{2}\Big[&\sin\frac{k\pi y_{j-2}}{y_3}\sinh^{-1}\frac{k\pi x_3}{y_3}\Big(-c^{(1)}_k - c^{(2)}_k \mathrm{e}^{-k\pi(x_3/y_3)}\Big)\\
&+\cos\frac{k\pi y_{j-2}}{y_3}\cosh^{-1}\frac{k\pi x_3}{y_3}\Big(c^{(3)}_k - c^{(4)}_k \mathrm{e}^{-k\pi(x_3/y_3)}\Big)\\
&+y_{j-2}\cos\frac{k\pi y_{j-2}}{y_3}\cosh^{-1}\frac{k\pi x_3}{y_3}\Big(a^{(4)}_k - b^{(4)}_k \mathrm{e}^{-k\pi(x_3/y_3)}\Big)\Big],
\end{aligned}$$

$$a^{[2]}_{jk,1} = \frac{1}{2}\sin\frac{k\pi y_{j-2}}{y_3}\sinh^{-1}\frac{k\pi x_3}{y_3}\Big(-a^{(4)}_k + b^{(4)}_k \mathrm{e}^{-k\pi(x_3/y_3)}\Big),$$

$$\begin{aligned}
a^{[3]}_{jk,0} = &\frac{1}{2\mathrm{i}}\sinh^{-1}\frac{k\pi y_3}{x_3}\Big(a^{(1)}_k\sinh\frac{k\pi y_{j-2}}{x_3} + b^{(1)}_k\sinh\frac{k\pi(y_3-y_{j-2})}{y_3}\Big)\\
&+\frac{1}{2}\cosh^{-1}\frac{k\pi y_3}{x_3}\Big(a^{(2)}_k\cosh\frac{k\pi y_{j-2}}{x_3} + b^{(2)}_k\cosh\frac{k\pi(y_3-y_{j-2})}{y_3}\Big)\\
&+y_{j-2}\Big(a^{(3)}_k\cosh\frac{k\pi y_{j-2}}{x_3} + b^{(3)}_k\cosh\frac{k\pi(y_3-y_{j-2})}{y_3}\Big),
\end{aligned}$$

$$a^{[3]}_{jk,1} = \frac{1}{2\mathrm{i}}\sinh^{-1}\frac{k\pi y_3}{x_3}\Big(a^{(3)}_k\sinh\frac{k\pi y_3}{x_3} + b^{(3)}_k\sinh\frac{k\pi(y_3-y_{j-2})}{x_3}\Big),$$

$$\begin{aligned}
a^{[4]}_{jk,0} = &-\frac{1}{2\mathrm{i}}\sinh^{-1}\frac{k\pi y_3}{x_3}\Big(a^{(1)}_k\sinh\frac{k\pi y_{j-2}}{x_3} + b^{(2)}_k\sinh\frac{k\pi(y_3-y_{j-2})}{y_3}\Big)\\
&+\frac{1}{2}\cosh^{-1}\frac{k\pi y_3}{x_3}\Big(a^{(2)}_k\cosh\frac{k\pi y_{j-2}}{x_3} + b^{(2)}_k\cosh\frac{k\pi(y_3-y_{j-2})}{x_3}\Big)\\
&+y_{j-2}\Big(a^{(3)}_k\cosh\frac{k\pi y_{j-2}}{x_3} + b^{(3)}_k\cosh\frac{k\pi(y_3-y_{j-2})}{y_3}\Big),
\end{aligned}$$

$$a^{[4]}_{jk,1} = -\frac{1}{2\mathrm{i}}\sinh^{-1}\frac{k\pi y_3}{x_3}\Big(a^{(3)}_k\sinh\frac{k\pi y_{j-2}}{x_3} + b^{(3)}_k\sinh\frac{k\pi(y_3-y_{j-2})}{y_3}\Big); \tag{3.2.13}$$

$$\psi_j(\zeta) = b^{[0]}_{j1,0} + b^{[0]}_{j1,1}\zeta + b^{[0]}_{j1,2}\zeta^2$$

$$+\sum_{k=1}^{\infty}\Big[b^{[1]}_{jk,0}\mathrm{e}^{k\pi(\zeta/x_3)}+b^{[1]}_{jk,1}\zeta\mathrm{e}^{k\pi(\zeta/x_3)}+b^{[2]}_{jk,0}\mathrm{e}^{-k\pi(\zeta/x_3)} + b^{[2]}_{jk,1}\zeta\mathrm{e}^{-k\pi(\zeta/x_3)}+b^{[3]}_{jk,0}\mathrm{e}^{\mathrm{i}k\pi(\zeta/y_3)}+b^{[4]}_{jk,0}\mathrm{e}^{-\mathrm{i}k\pi(\zeta/y_3)} + b^{[3]}_{jk,1}\zeta\mathrm{e}^{\mathrm{i}k\pi(\zeta/y_3)}+b^{[4]}_{jk,1}\zeta\mathrm{e}^{-\mathrm{i}k\pi(\zeta/y_3)}\Big],$$

$$b^{[0]}_{j1,0}=a^{(0)}_0+a^{(1)}_0x_{j-2}+a^{(2)}_0x^2_{j-2},$$

$$b^{[0]}_{j1,1}=b^{(1)}_0,\qquad b^{[0]}_{j1,2}=b^{(2)}_0,$$

$$b^{[1]}_{jk,0}=-\frac{1}{2}\cos\frac{k\pi x_{j-2}}{x_3}\sinh^{-1}\frac{k\pi y_3}{x_3}\Big(a^{(1)}_k-b^{(1)}_k\mathrm{e}^{-k\pi(y_3/x_3)}\Big) +\frac{1}{2}\cos\frac{k\pi x_{j-2}}{x_3}\cosh^{-1}\frac{k\pi y_3}{x_3}\Big(a^{(2)}_k-b^{(2)}_k\mathrm{e}^{-k\pi(y_3/x_3)}\Big) +\frac{x_{j-2}}{2}\cos\frac{k\pi x_{j-2}}{x_3}\cosh^{-1}\frac{k\pi y_3}{x_3}\Big(a^{(3)}_k-b^{(3)}_k\mathrm{e}^{-k\pi(y_3/x_3)}\Big),$$

$$b^{[1]}_{jk,1}=\frac{1}{2}\sin\frac{k\pi x_{j-2}}{x_3}\sinh^{-1}\frac{k\pi y_3}{x_3}\Big(a^{(3)}_k-b^{(3)}_k\mathrm{e}^{-k\pi(y_3/x_3)}\Big),$$

$$b^{[2]}_{jk,0}=\frac{1}{2}\Big[\sin\frac{k\pi x_{j-2}}{x_3}\sinh^{-1}\frac{k\pi y_3}{x_3}\Big(-a^{(3)}_k-b^{(1)}_k\mathrm{e}^{-k\pi(y_3/x_3)}\Big) +\frac{1}{2}\cos\frac{k\pi x_{j-2}}{x_3}\cosh^{-1}\frac{k\pi y_3}{x_3}\Big(a^{(2)}_k+b^{(2)}_k\mathrm{e}^{-k\pi(y_3/x_3)}\Big) +\frac{x_{j-2}}{2}\cos\frac{k\pi x_{j-2}}{x_3}\cosh^{-1}\frac{k\pi y_3}{x_3}\Big(-a^{(3)}_k-b^{(3)}_k\mathrm{e}^{-k\pi(y_3/x_3)}\Big)\Big],$$

$$b^{[2]}_{jk,1}=\frac{1}{2}\sin\frac{k\pi x_{j-2}}{x_3}\sinh^{-1}\frac{k\pi y_3}{x_3}\Big(-a^{(3)}_k+b^{(3)}_k\mathrm{e}^{-k\pi(y_3/x_3)}\Big),$$

$$b^{[3]}_{jk,0}=\frac{1}{2\mathrm{i}}\sinh^{-1}\frac{k\pi x_3}{y_3}\Big(c^{(1)}_k\sinh\frac{k\pi x_{j-2}}{y_3}+c^{(2)}_k\sinh\frac{k\pi(x_3-x_{j-2})}{y_3}\Big) +\frac{1}{2}\cosh^{-1}\frac{k\pi x_3}{y_3}\Big(c^{(3)}_k\sinh\frac{k\pi x_{j-2}}{y_3}+c^{(4)}_k\sinh\frac{k\pi(x_3-x_{j-2})}{y_3}\Big) +\frac{1}{2\mathrm{i}}\sinh^{-1}\frac{k\pi x_3}{y_3}\Big(a^{(3)}_k\sinh\frac{k\pi x_{j-2}}{y_3}+b^{(4)}_k\sinh\frac{k\pi(x_3-x_{j-2})}{y_3}\Big),$$

$$b^{[3]}_{jk,1}=\frac{1}{2\mathrm{i}}\cosh^{-1}\frac{k\pi x_3}{y_3}\Big(a^{(4)}_k\cosh\frac{k\pi x_{j-2}}{y_3}+b^{(4)}_k\cosh\frac{k\pi(x_3-x_{j-2})}{y_3}\Big),$$

$$b^{[4]}_{jk,0}=-\frac{1}{2\mathrm{i}}\sinh^{-1}\frac{k\pi x_3}{y_3}\Big(c^{(1)}_k\sinh\frac{k\pi x_{j-2}}{y_3}+c^{(2)}_k\sinh\frac{k\pi(x_3-x_{j-2})}{y_3}\Big) +\frac{1}{2}\cosh^{-1}\frac{k\pi x_3}{y_3}\Big(c^{(3)}_k\cosh\frac{k\pi x_{j-2}}{y_3}+c^{(4)}_k\cosh\frac{k\pi(x_3-x_{j-2})}{y_3}\Big)$$

$$+\frac{x_{j-2}}{4}\sinh^{-1}\frac{k\pi x_3}{y_3}\left(a_k^{(4)}\sinh\frac{k\pi x_{j-2}}{y_3}+b_k^{(4)}\sinh\frac{k\pi(x_3-x_{j-2})}{y_3}\right),$$

$$b_{jk,1}^{[4]}=\frac{1}{2\mathrm{i}}\cosh^{-1}\frac{k\pi x_3}{y_3}\left(a_k^{(4)}\cosh\frac{k\pi x_{j-2}}{y_3}+b_k^{(4)}\cosh\frac{k\pi(x_3-x_{j-2})}{y_3}\right). \tag{3.2.14}$$

Since $\varphi_j(z)=0$, $j=3,4$, on the real segment $[x_1,x_2]$ and $\psi_j(\zeta)=0$, $j=3,4$, on the real segment $[y_1,y_2]$, therefore, by uniqueness of continuation of analytic functions of one complex variable, we have $\varphi_j(z)=0$, $j=3,4$, everywhere in $\overline{P}$ and $\psi_j(\zeta)=0$, $j=3,4$, everywhere in $\overline{\Pi}$.

Introduce the function

$$Q_1(\lambda)=\sin^2\lambda x_3\sinh\lambda y_3.$$

For the function $Q_1(\lambda)$, the Borel functional transform

$$\gamma(z)=\frac{1}{z-(y_3+2x_3\mathrm{i})}+\frac{1}{z-(y_3+2x_3\mathrm{i})}+\frac{1}{z-(y_3-2x_3\mathrm{i})}-\frac{1}{z+(y_3+2x_3\mathrm{i})}-\frac{2}{z-y_3}+\frac{2}{z+y_3}$$

has singularities only at the points

$$y_3+2x_3\mathrm{i},\quad -y_3+2x_3\mathrm{i},\quad y_3-2x_3\mathrm{i},\quad -y_3-2x_3\mathrm{i},\quad y_3,\quad -y_3.$$

Therefore, the conjugate diagram of $Q_1(\lambda)$ is

$$\overline{P_1}=\{-y_3\le \operatorname{Re}z\le y_3,\ -2x_3\le \operatorname{Im}z\le 2x_3\}.$$

By assumption of the theorem, $\overline{P_1}\subset P$.

Let C_1 be a simple closed contour lying in $P\setminus\overline{P_1}$.

We shall denote by $A_{k,m}^{[i]}(t)$ the Borel functional transform of the entire function

$$\frac{Q_1(\lambda)}{(\lambda-\lambda_k^{[i]})^{m+1}},\qquad m=0,1,2,\dots,\nu_k^{[i]}-1,\qquad i=\overline{1,4},$$

where $\lambda_k^{[i]}$ are zeros of $Q_1(\lambda)$ with multiplicities $\nu_k^{[i]}$.

We set

$$M_{1k}^{[i]}(\mu)=(\mu-\lambda_k^{[i]})^{\nu_k^{[i]}}Q_1^{-1}(\mu),$$

$$\varphi^{[i]}_{k,m}(t) = \frac{1}{m!\,(\nu^{[i]}_k - m - 1)!}$$
$$\times \sum_{q=0}^{\nu^{[i]}_k - m - 1} q! \cdot C^q_{\nu^{[i]}_k - m - 1} [M^{[i]}_{1k}(\mu)]^{(\nu^{[i]}_k - m - q - 1)}_{\mu = \lambda^{[i]}_k} A^{[i]}_{k,q}(t).$$

Then

$$a^{[i]}_{jk,1} = -\frac{1}{2\pi \mathrm{i}} \int_{C_1} \varphi_j(z) \varphi^{[i]}_{k,1}(z)\,\mathrm{d}z, \quad j = 3,4, \quad k = 1,2\dots. \tag{3.2.15}$$

By analogous considerations we find

$$b^{[4]}_{jk,1} = \frac{1}{2\pi \mathrm{i}} \int_{C_2} \psi_j(\zeta) \psi^{[4]}_{k,1}(\zeta)\,\mathrm{d}\zeta, \quad j = 3,4, \quad k = 1,2,\dots, \tag{3.2.16}$$

where C_2 is a simple closed contour lying in $\Pi \setminus \overline{\Pi_1}$ ($\overline{\Pi_1} \subset \Pi$), and

$$\overline{\Pi_1} = \{-x_3 \le \operatorname{Re}\zeta \le x_3, \ -2x_3 \le \operatorname{Im}\zeta \le 2x_3\}.$$

We recall now that $\varphi_j(z) = 0$, $j = 3,4$, in P and $\psi_j(\zeta) = 0$, $j = 3,4$, in Π. Then from (3.2.15) and (3.2.16) we have

$$a^{[4]}_{jk,1} = b^{[4]}_{jk,1} = 0, \qquad j = 3,4, \quad k = 1,2,\dots,$$

which together with (3.2.9) and (3.2.10) implies that all the constants in (3.2.6) are equal to zero. Therefore,

$$u(x,y) = 0 \qquad \text{in} \quad D_1.$$

□

Note that a similar result can be obtained by applying the Jensen inequality. Note also that if instead of (3.2.6) we take the function

$$u(x,y) = a^{(3)}_0 + b^{(4)}_0 x + b^{(3)}_0 y + c^{(4)}_0 x^2 + c^{(3)}_0 y^2 + d^{(4)}_0 x^3 + d^{(3)}_0 y^3$$
$$+ \sum_{k=1}^{\infty} \Big[a^{(1)}_k \cosh\frac{k\pi y}{x_3} + b^{(1)}_k \sinh\frac{k\pi y}{x_3}$$
$$+ y\Big(c^{(1)}_k \cosh\frac{k\pi y}{x_3} + d^{(1)}_k \sinh\frac{k\pi y}{x_3}\Big)\Big] \sin\frac{k\pi x}{x_3}$$
$$+ \sum_{k=1}^{\infty} \Big[a^{(2)}_k \cosh\frac{k\pi x}{y_3} + b^{(2)}_k \sinh\frac{k\pi x}{y_3}$$

$$+ x\left(c_k^{(2)} \cosh \frac{k\pi x}{y_3} + d_k^{(2)} \sinh \frac{k\pi x}{y_3}\right)\Big] \sin \frac{k\pi y}{y_3}$$

$$+ \sum_{k=1}^{\infty} \Big[a_k^{(3)} \cosh \frac{k\pi y}{x_3} + b_k^{(3)} \sinh \frac{k\pi y}{x_3}$$

$$+ y\left(c_k^{(3)} \cosh \frac{k\pi y}{x_3} + d_k^{(3)} \sinh \frac{k\pi y}{x_3}\right)\Big] \cos \frac{k\pi x}{x_3}$$

$$+ \sum_{k=1}^{\infty} \Big[a_k^{(4)} \cosh \frac{k\pi x}{y_3} + b_k^{(4)} \sinh \frac{k\pi x}{y_3}$$

$$+ x\left(c_k^{(4)} \cosh \frac{k\pi x}{y_3} + d_k^{(4)} \sinh \frac{k\pi x}{y_3}\right)\Big] \cos \frac{k\pi y}{y_3},$$

then (3.2.7) should be replaced by the inequality

$$|a_k^{(i)}|,\ |b_k^{(i)}|,\ |c_k^{(i)}|,\ |d_k^{(i)}| \le A\mathrm{e}^{-kd}, \qquad i = \overline{1,4}, \qquad k = 1, 2, \dots,$$

with appropriate choice

$$A > 0, \qquad d > 2\max\left\{\frac{x_3}{y_3}, \frac{y_3}{x_3}\right\}\pi.$$

Chapter 4.

Internal boundary value problems and the Cauchy problem for the abstract biharmonic equation

4.1. INTERNAL BOUNDARY VALUE PROBLEMS

1. Let $x(t)$ be a function of the scalar argument $t \in [0, \infty)$ with values in a Hilbert space H. Let A be a linear unbounded constant self-adjoint positive definite operator acting from H to H. Besides, we suppose that A has a purely point spectrum and the domain of definition $D(A^2)$ is everywhere dense in H.

We consider the homogeneous equation

$$\left(\frac{\mathrm{d}^2}{\mathrm{d}t^2} - A\right)^2 x(t) = 0, \qquad 0 < t < \infty. \tag{4.1.1}$$

We assume that the operator A is given. We need to find the function $x(t)$ satisfying the conditions

$$x(0) = x_0, \quad x(T) = x_T \quad (0 < T < \infty), \quad x(\infty) = 0, \tag{4.1.2}$$

where $x_0, x_T \in H$.

The conditions imposed on the operator A imply that the operator A has a complete orthonormal system of eigenfunctions $\{\varphi_k\}$ and the corresponding system of positive eigenvalues $\{\lambda_k^2\}$. Then we have

$$x(t) = \sum_{k=1}^{\infty} x_k(t)\varphi_k, \qquad x_k = (x, \varphi_k), \tag{4.1.3}$$

so, (4.1.1) decomposes into the system of ordinary differential equations

$$\Big(\frac{\mathrm{d}^2}{\mathrm{d}t^2} - \lambda_k^2\Big)^2 x_k(t) = 0.$$

Its solution is as follows

$$x_k(t) = \Big(\mathrm{e}^{-\lambda_k t}\frac{\lambda_k}{T}\mathrm{e}^{-\lambda_k T}\Big)x_{0k} + \frac{t}{T}\mathrm{e}^{-\lambda_k(t-T)}x_{Tk},$$

where

$$x_{0k} = (x_0, \varphi_k), \qquad x_{Tk} = (x_T, \varphi_k).$$

This representation implies that $0 < t < T$ and the problem (4.1.1), (4.1.2) is classically ill-posed (Lavrent'ev, 1973).

Note that the problem (4.1.1), (4.1.2) includes the problem for the biharmonic equation considered in Section 1.3.

Theorem 4.1.1. *For each solution of equation* (4.1.1) *the following inequality holds true*

$$\|x(t)\| \le \sqrt{2}\,\big(\mathrm{e}^{-\lambda_1 t}\|x(0)\|^2 + t^2\|x_2(t)\|^2\big)^{1/2}, \tag{4.1.4}$$

where

$$\|x_2(t)\| \le \begin{cases} \dfrac{\sqrt{2}}{T}\big(\|x(T)\|^2 + \|x(0)\|^2\big)^{t/2T}\|x''(T) - Ax(T)\|^{(T-t)/T}, \\ \hfill 0 \le t \le T, \\ \dfrac{\sqrt{2}}{T}\big(\|x(T)\|^2 + \|x(0)\|^2\big)^{t/2}\mathrm{e}^{-\lambda_1(t-T)}, \qquad T \le t \le \infty, \end{cases}$$

or

$$\|x_2(t)\| \le \begin{cases} \dfrac{(t-T)(x_2'(T), x_2(T))}{2\|x_2(T)\|^2}, & 0 \le t \le T, \\ \|x(T)\|^2\mathrm{e}^{-\lambda_1(t-T)}, & T \le t \le \infty. \end{cases}$$

Here

$$x_2(T) = \frac{1}{T}(x(T) - x_1(T)),$$

$$x_2'(T) = x''(T) - Ax(T)$$
$$= \frac{1}{T}x'(T) - \frac{1}{T^2}x(T) - \frac{1}{T^2}x_1(T) - \frac{1}{T}x'(T);$$

$x_1(t)$ is the solution of the problem

$$x_1''(t) - Ax_1(t) = 0, \qquad x_1(0) = x(0);$$

and λ_1 is the smallest eigenvalue of the operator A.

Note that inequality (4.1.4) implies uniqueness of solution of the problem in the class of functions with bounded first and second derivatives.

The proof is based on the representation

$$x(t) = x_1(t) + t\,x_2(t), \qquad x_2''(t) - Ax_2(t) = 0$$

of any solution of equation (4.1.1).

Consideration of the ill-posed (in the classical sense) problem (4.1.1), (4.1.2) as a conditionally well-posed one allows us to construct an approximate solution. Given approximate data $x_{0\varepsilon}$ and $x_{T\varepsilon}$, $\|x_{0\varepsilon} - x_0\| \le \varepsilon$, $\|x_{T\varepsilon} - x_T\| \le \varepsilon$, we construct an approximate solution with guaranteed accuracy by the regularization method (Lavrent'ev, 1973).

Indeed, if

$$x_{n\varepsilon}(t) = \sum_{k=1}^{n}\Big(\mathrm{e}^{-\lambda_k t}\frac{t}{T}\mathrm{e}^{-\lambda_k T}\Big)x_{0k\varepsilon}\varphi_k + \frac{t}{T}\sum_{k=1}^{n}\mathrm{e}^{-\lambda_k(t-T)}x_{Tk\varepsilon}\varphi_k,$$

where

$$x_{0k\varepsilon} = (x_{0\varepsilon}, \varphi_k), \qquad x_{Tk\varepsilon} = (x_{T\varepsilon}, \varphi_k),$$

then under the condition $\|x''(0) - Ax_2(0)\| \le M$ we have

$$\|x - x_{n\varepsilon}\| \le \sqrt{2}\,\frac{\mathrm{e}^{\lambda_n T}}{\lambda_n T}\,\varepsilon + \big(2\|x_0\|^2 + t^2\mathrm{e}^{-2\lambda_{n+1}t}M\big)^{1/2} = \omega(n,\varepsilon).$$

For fixed accuracy ε of the approximate data, the value of the parameter n for which $\inf\omega(n,\varepsilon)$ is achieved will be optimal in the sense of the above estimate.

Example 4.1.1. If

$$A = -\sum_{i=1}^{n}\frac{\partial^2}{\partial x_i^2} \qquad (0 \le x_i \le a_i),$$

then

$$\lambda_{\mathbf{m}}^2 = \lambda_{m_1 \cdots m_n}^2 = \pi \sum_{i=1}^{n} \frac{m_i^2}{a_i^2},$$

$$\varphi_{\mathbf{m}}(x) = \varphi_{\mathbf{m}}(x_1, \dots, x_n) = 2^{n/2} \prod_{i=1}^{n} a_i^{-1/2} \sin \frac{m_i \pi x_i}{a_i},$$

$$\Big(\frac{\mathrm{d}^2}{\mathrm{d}t^2} - A\Big)^2 x = \Big(\frac{\mathrm{d}^2}{\mathrm{d}t^2} + \sum_{i=1}^{m} \frac{\partial^2}{\partial x_i^2}\Big)^2 x = 0.$$

2. Let $x(t)$ be a function of a scalar argument $t \in [0, 1]$ with values in a Hilbert space H, and let A be a linear unbounded constant self-adjoint positive definite operator acting from H to H. We also assume that A has a purely point spectrum and the domain of definition $D(A^2)$ is everywhere dense in H.

Consider the homogeneous equation

$$\Big(\frac{\mathrm{d}^2}{\mathrm{d}t^2} - A\Big)^2 x(t) = 0, \qquad 0 < t < 1.$$

We assume that the operator A is given and we need to find the function $x(t)$ satisfying the conditions

$$x(t_i) = 0, \qquad i = 0, 1, 2, \qquad t_0 = 0.$$

Theorem 4.1.2. *If the function $x(t)$ satisfies the inequalities*

$$\|x(t_3)\| \le \varepsilon, \qquad \|x_t'(t_3)\| \le M,$$

then the following inequality holds

$$\|x(t)\| \le \frac{A_\nu(t)}{A_\nu(t_3)}\, \varepsilon,$$

where

$$\begin{aligned} A_\nu(t) = {} & \frac{1}{t_1 t_2 \sinh \lambda_\nu (t_2 - t_1)} \Big\{ t_1 t_2 \sinh \lambda_\nu (t_2 - t_1) \sinh \lambda_\nu t \\ & + t \big[(t_1 - t_2) \sinh \lambda_\nu t_1 \, \sinh \lambda_\nu t_2 \, \cosh \lambda_\nu t \\ & + (t_2 \sinh \lambda_\nu t_1 \cosh \lambda_\nu t_2 - t_1 \cosh \lambda_\nu t_1 - \sinh \lambda_\nu t_2) \sinh \lambda_\nu t \big] \Big\}; \end{aligned} \qquad (4.1.5)$$

ν is the positive root of the equation

$$\left|\frac{A'_{\nu t}(t_3)}{A_\nu(t_3)}\right| = \frac{M}{\varepsilon}.$$

Proof. It is easy to verify that the function

$$x(t) = \sum_{k=1}^{\infty} b_k A_k(t)\varphi_k \tag{4.1.6}$$

satisfies equation (4.1.1).

From (4.1.3) and (4.1.6) we get

$$\sum_{k=1}^{\infty} b_k^2 A_k^2(t_3) \le \varepsilon^2, \qquad \sum_{k=1}^{\infty} b_k^2 A_k'^2(t_3) \le M^2. \tag{4.1.7}$$

Taking into account (4.1.7), we estimate the function

$$\omega(t) = \sum_{k=1}^{\infty} b_k^2 A_k^2(t) = \|x(t)\|^2. \tag{4.1.8}$$

By (4.1.7), this sum is bounded.

We suppose that the sum (4.1.8) achieves its conditional maximum relative to the variables b_k for $b_k = \bar{b}_k$, $k = 1, 2, \ldots$. To find this conditional maximum we use the Lagrange method.

Let us find stationary points of the functional

$$F(t, b_k) = \sum_{k=1}^{\infty} b_k^2 \big[A_k^2(t) + \lambda A_k^2(t_3) + \mu A_{kt}'^2(t_3)\big],$$

where λ and μ are the Lagrange multipliers.

Equating the linear part of the increment $F(t, b_k + a_k) - F(t, b_k)$ to zero, we obtain

$$2\bar{b}_k \big[A_k^2(t) + \lambda A_k^2(t_3) + \mu A_{kt}'^2(t_3)\big] = 0,$$

which implies

$$\bar{b}_k = 0, \qquad k \ne p, q,$$

where p and q ($p < q$) are some numbers.

Now, by (4.1.7), $\bar{b}_p$ and $\bar{b}_q$ satisfy one of the following three relations

$$\begin{aligned}
&\text{(I)} \qquad \begin{cases} \bar{b}_p^2 A_p^2(t_3) + \bar{b}_q^2 A_q^2(t_3) = \varepsilon^2, \\ \bar{b}_p^2 A_{pt}'^2(t_3) + \bar{b}_q^2 A_{qt}'^2(t_3) = M^2, \end{cases} \\
&\text{(II)} \qquad \bar{b}_p = 0, \\
&\text{(III)} \qquad \bar{b}_q = 0.
\end{aligned} \tag{4.1.9}$$

We suppose first that relations (I) hold. Then

$$\begin{aligned}
\bar{b}_p^2 &= \frac{\varepsilon^2 A_{qt}'^2(t_3) - M^2 A_q^2(t_3)}{A_p^2(t_3) A_{qt}'^2(t_3) - A_q^2(t_3) A_{pt}'^2(t_3)}, \\
\bar{b}_q^2 &= \frac{M^2 A_p^2(t_3) - \varepsilon^2 A_{pt}'^2(t_3)}{A_p^2(t_3) A_{qt}'^2(t_3) - A_q^2(t_3) A_{pt}'^2(t_3)}.
\end{aligned} \tag{4.1.10}$$

Note that (4.1.10) implies

$$\frac{A_{qt}'(t_3)}{A_q(t_3)} > \frac{M}{\varepsilon} > \frac{A_{pt}'(t_3)}{A_p(t_3)}.$$

Taking into account (4.1.9) and (4.1.10), we obtain

$$\begin{aligned}
\sum_{k=1}^{\infty} \bar{b}_k^2 A_k^2(t) = \varepsilon^2 & \frac{A_{qt}'^2(t_3) A_p^2(t) - A_{pt}'^2(t_3) A_q^2(t)}{A_p^2(t_3) A_{qt}'^2(t_3) - A_q^2(t_3) A_{pt}'^2(t_3)} \\
&+ M^2 \frac{A_p^2(t_3) A_q^2(t) - A_q^2(t_3) A_p^2(t)}{A_p^2(t_3) A_{qt}'^2(t_3) - A_q^2(t_3) A_{pt}'^2(t_3)}.
\end{aligned} \tag{4.1.11}$$

It is easy to show that the following functions of p and q $(p < q)$

$$\begin{aligned}
F_1(p, q) &= \frac{A_{qt}'^2(t_3) A_p^2(t) - A_{pt}'^2(t_3) A_q^2(t)}{A_p^2(t_3) A_{qt}'^2(t_3) - A_q^2(t_3) A_{pt}'^2(t_3)}, \\
F_2(p, q) &= \frac{-A_q^2(t_3) A_p^2(t) + A_p^2(t_3) A_q^2(t)}{A_p^2(t_3) A_{qt}'^2(t_3) - A_q^2(t_3) A_{pt}'^2(t_3)}
\end{aligned} \tag{4.1.12}$$

increase when p increases and q decreases. Hence, taking into account (4.1.11) and (4.1.12), we get

$$\sum_{k=1}^{\infty} \bar{b}_k^2 A_k^2(t) \le \lim_{p-q \to 0} [\varepsilon^2 F_1(p, q) + M^2 F_2(p, q)] = \frac{A_\nu^2(t)}{A_\nu^2(t_3)} \varepsilon^2. \tag{4.1.13}$$

Now we suppose that relation (II) from (4.1.9) holds. Then

$$\sum_{k=1}^{\infty} \bar{b}_k^2 A_k^2(t) = \bar{b}_q^2 A_q^2(t)$$

and, by (4.1.7), we have

$$\bar{b}_q^2 A_q^2(t_3) \leq \varepsilon^2, \qquad \bar{b}_q^2 A_{qt}'^2(t_3) \leq M^2. \tag{4.1.14}$$

Hence we obtain

$$\sum_{k=1}^{\infty} \bar{b}_k^2 A_k^2(t) \leq \frac{A_\nu^2(t)}{A_\nu^2(t_3)} \varepsilon^2. \tag{4.1.15}$$

In the case where (III) holds, inequality (4.1.15) is obtained analogously.

The statement of the theorem follows from (4.1.13), (4.1.14), and from the definition of the numbers $\bar{b}_k$. □

For regularization of the problem (4.1.1), (4.1.2) we consider the family of linear operators B_n depending on an integer parameter and defined as follows

$$B_n x(t) = \sum_{k=1}^{n} b_k A_k(t) \varphi_k. \tag{4.1.16}$$

It is easy to see that the family of operators B_n will be regularizing with respect to our problem if we consider the data

$$x(t_i) = \sum_{k=1}^{\infty} b_k A_k(t_i) \varphi_k, \qquad i = 0, 1, 2,$$

and the solution $x(t)$ as elements of the Hilbert spaces H (see Lavrent'ev, 1973, p. 57). From (4.1.6), (4.1.16), and (4.1.7) it follows that

$$\|B_n x(t) - x(t)\|^2 = \sum_{k=n+1}^{\infty} b_k^2 A_k^2(t), \tag{4.1.17}$$

$$\sum_{k=1}^{\infty} b_k^2 A_{kt}'^2(t_3) \leq M^2. \tag{4.1.18}$$

We see that the sum on the right-hand side of (4.1.17) under condition (4.1.18) attains its maximal value in the case where

$$b_k = 0, \qquad k \neq n+1,$$

$$b_{n+1}^2 = [A'_{(n+1)t}(t_3)]^{-2} M^2.$$

Therefore,

$$\|B_n x(t) - x(t)\| \le \frac{A_{n+1}(t)}{A'_{(n+1)t}(t_3)} M,$$

$$\|B_n\| = A_n(t).$$

Assume that we know an element $x_\varepsilon(t_i)$ such that $\|x(t_3) - x_\varepsilon(t_i)\| \le \varepsilon$. Then

$$\|B_n x(t) - x(t)\| \le A_n(t)\varepsilon + \frac{A_{n+1}(t)}{A'_{(n+1)t}(t_3)} M. \qquad (4.1.19)$$

For fixed accuracy ε of approximate data the value of the parameter n for which the infimum

$$\inf \left\{ A_n(t)\varepsilon + \frac{A_{n+1}(t)}{A'_{(n+1)t}(t_3)} M \right\}$$

is achieved will be optimal in the sense of the estimate (4.1.19).

4.2. THE CAUCHY PROBLEM

We consider the problem

$$Lu \equiv \left(\frac{\mathrm{d}^2}{\mathrm{d}t^2} + A\right)^2 u(t) = 0, \qquad 0 < t < T < \infty, \qquad (4.2.1)$$

$$u^{(i)}(t)|_{t=0} = u_0^i, \qquad i = 0, 1, 2, 3. \qquad (4.2.2)$$

Here $u(t)$ is a function of the scalar argument t with values in a real Hilbert space H; A is a linear unbounded normal operator with a dense domain of definition $D(A^2) \subset H$. In particular,

$$A = \frac{\partial^2}{\partial x^2} \qquad (0 < x < \infty).$$

The problem (4.2.1), (4.2.2) is essentially ill-posed (Tikhonov and Arsenin, 1977). Let us investigate this problem for conditional well-posedness (Lavrent'ev, 1973).

Theorem 4.2.1. *Let $u \in L_2(H,[0,T])$ and*

$$\|A^k u^{(i)}(t)\| < M_{ki}, \qquad k,i = 0,1,2,3. \tag{4.2.3}$$

Then, for each solution of the problem (4.2.1), (4.2.2) on the segment $[0,T]$ the following inequality holds

$$\int_\sigma^T \|u(t)\|_H^2 \,\mathrm{d}t \le \frac{2K(T)}{\sigma^{14}}, \tag{4.2.4}$$

where σ^2 is the root of the equation

$$\frac{K(T)}{\sigma^2} = K(0)\mathrm{e}^{\sigma^2 T^2/2} \tag{4.2.5}$$

and

$$K(t) \equiv K(\|A^k u^{(i)}(t)\|_H, \|A^{2i} u^{(j)}(t)\|), \qquad i,j = \overline{0,3}, \quad l = \overline{0,2},$$

is a quadratic form with coefficients depending only on t, T, and $M = \max M_{ki}$.

Evidently, $\sigma^2 \to \infty$ for $\|u_0^{(i)}\| \to 0$, $i = \overline{0,3}$.

Proof. To prove this theorem we use the exponential method. Let

$$u_\sigma = \mathrm{e}^{\sigma^2(t-T)^2/2} u. \tag{4.2.6}$$

Hence, substituting u and its derivatives in (4.2.1), we have

$$\begin{aligned} L_\sigma u_\sigma \equiv{}& \frac{\mathrm{d}^4 u_\sigma}{\mathrm{d}t^4} + [-4\sigma^2(t-T)]\frac{\mathrm{d}^3 u_\sigma}{\mathrm{d}t^3} + [-6\sigma^2 + 6\sigma^4(t-T)^2 + 2A]\frac{\mathrm{d}^2 u_\sigma}{\mathrm{d}t^2} \\ &+ [-4A\sigma^2(t-T) + 12\sigma^4(t-T) - 4\sigma^6(t-T)^3]\frac{\mathrm{d}u_\sigma}{\mathrm{d}t} \\ &+ [A^2 - 2A\sigma^2 + 2A\sigma^4(t-T)^2 + 3\sigma^4 - 6\sigma^6(t-T)^2 \\ &\quad + \sigma^8(t-T)^4]u_\sigma = 0, \end{aligned} \tag{4.2.7}$$

$$\begin{aligned} L_\sigma^+ u_\sigma \equiv{}& \frac{\mathrm{d}^4 u_\sigma}{\mathrm{d}t^4} + [-4\sigma^2(t-T)]\frac{\mathrm{d}^3 u_\sigma}{\mathrm{d}t^3} + [2A - 18\sigma^2 + 6\sigma^4(t-T)^2]\frac{\mathrm{d}^2 u_\sigma}{\mathrm{d}t^2} \\ &+ [-4A\sigma^2 + 12\sigma^4 + 24\sigma^4(t-T) - 12\sigma^6(t-T)^2]\frac{\mathrm{d}u_\sigma}{\mathrm{d}t} \\ &+ [-6A\sigma^2 + 15\sigma^4 + 2A\sigma^4(t-T)^2 \\ &\quad - 18\sigma^6(t-T)^2 + \sigma^8(t-T)^4 + A^2]u_\sigma, \end{aligned} \tag{4.2.8}$$

where L_σ^+ is the operator conjugate to L_σ in the Lagrange sense. Now, taking into account (4.2.7) and (4.2.8), we find

$$(L_\sigma L_\sigma^+ - L_\sigma^+ L_\sigma)u_\sigma = -R_\sigma u_\sigma = \sum_{i=0}^{6} \omega_i \frac{\mathrm{d}^{6-i} u_\sigma}{\mathrm{d}t^{6-i}}, \tag{4.2.9}$$

where $\omega_0, \omega_1, \dots, \omega_6$ are certain operators which can be written explicitly.

Multiply both sides of equality (4.2.9) scalarly by u_σ and integrate for t going from 0 to T. Further, applying the formula of integration by parts, we obtain the following equality

$$\sigma^{14} \int_0^T (t-T)^6 \|u_\sigma\|^2 \,\mathrm{d}t - \int_0^T (Pu_\sigma, u_\sigma)\,\mathrm{d}t + \int_0^T (Qu'_\sigma, u'_\sigma)\,\mathrm{d}t - 32\sigma^2 \|u'''_\sigma\|^2$$
$$= [M_1(t) - M_2(t)]\big|_0^T + \int_0^T \|L_\sigma^+ u_\sigma\|^2 \,\mathrm{d}t, \tag{4.2.10}$$

where $M_1(t)$ depends on scalar products of the form $([L_\sigma^+ u_\sigma]^{(i)}, u_\sigma^{(k)})$, $i = \overline{0,3}$; $M_2(t)$ depends on scalar products of the form $(u_\sigma^{(i)}(t), u_\sigma^{(k)}(t))$, $i = \overline{0,3}$, $k = \overline{0,2}$; and P and Q are certain operators which can be written explicitly.

Recalling condition (4.2.3), we apply a lemma from the book (Kostrikin, 1977, p. 273, Lemma 2). Then we reject the positive term in (4.2.10) and obtain

$$\sigma^{14} \int_0^T \|u_\sigma(t)\|^2 \,\mathrm{d}t \le [M_1(t) - M_2(t)]\big|_0^T. \tag{4.2.11}$$

From (4.2.7) we find the fourth derivative $u_\sigma^{(\mathrm{IV})}$ and differentiate it. Then we express $u_\sigma^{(\mathrm{V})}$ and $[L_\sigma^+ u_\sigma]^{(i)}$, $i = \overline{0,3}$, via $u_\sigma^{(i)}(t)$, $i = \overline{0,3}$. Then from inequality (4.2.11) we obtain

$$\sigma^{14} \int_0^T \|u_\sigma(t)\|^2 \,\mathrm{d}t \le M_3(t)\big|_0^T, \tag{4.2.12}$$

where $M_3(t)$ depends only on scalar products of the form

$$(A^i u_\sigma^{(k)}, A^{2l} u_\sigma^{(m)}), \qquad i, k, m = \overline{0,3}, \quad l = \overline{0,2}.$$

Returning to the function $u(t)$ and using the inequality

$$1 \le \mathrm{e}^{\sigma^2 (t-T)^2/2} \le \infty,$$

from (4.2.12) we find

$$\sigma^{14} \int_0^T \|u(t)\|^2 \, \mathrm{d}t \le K(T) + K(0)\sigma^2 \mathrm{e}^{\sigma^2 T^2/2}, \tag{4.2.13}$$

which yields inequality (4.2.4) of the theorem. □

Note that an analogous theorem can be established for an equation of m-th order ($m \ge 1$) with unbounded, normally permutable operator coefficients.

Chapter 5.

Appendices

5.1. APPENDIX A

Here we apply the method of linear transformation of equations (Krein, 1971, §10) in the class $C^2(\overline{D_1}) \cap C^4(D_1)$ to obtain one sufficient condition of uniqueness of solutions of the first internal boundary value problem and the problem conjugate to it for the biharmonic equation.

Let D_α $(\alpha = 1, 2)$ be finite domains of the Euclidean space $\mathbb{R}^m$ $(m \geq 2)$ bounded by the Lyapunov surfaces S_α $(\alpha = 1, 2)$, respectively, such that $D_1 \supset \overline{D_2}$.

Problem. We need to find a function $u(p)$, $p = (x_1, x_2, \ldots, x_m) \in D_1$, which satisfies the conditions

$$\Delta^2 u(p) = 0, \quad p \in D_1, \tag{5.1.1}$$

$$u(p) = \varphi_\alpha(p), \quad p \in S_\alpha \quad (\alpha = 1, 2). \tag{5.1.2}$$

Here Δ is the Laplace operator, and $\varphi_\alpha(p)$ $(\alpha = 1, 2)$ are continuous functions given on S_α $(\alpha = 1, 2)$, respectively.

It is easy to see that the problem (5.1.1), (5.1.2) can be reduced to the integral Fredholm equation of first kind

$$Gv(Q) = f(p), \quad Q \in D_1, \quad p \in S_2, \tag{5.1.3}$$

where

$$Gv = -\int_{D_1} G(p, Q) v(Q) \, \mathrm{d}Q, \qquad \Delta v(Q) = 0, \quad Q \in D_1,$$

$$f(p) = \varphi_2(p) - \int_{S_1} \varphi_1(Q) \frac{\partial G(p, Q)}{\partial n_Q} \, \mathrm{d}_Q S_1.$$

Here $G(p, Q)$ is the Green function of the Dirichlet problem for the Laplace equation.

Theorem 5.1.1. *The problem* (5.1.1), (5.1.2) *has a unique solution in the class* $C^2(\overline{D_1}) \cap C^4(D_1)$ *if none eigenfunction of the Laplace operator of the Dirichlet problem in* D_1 *vanishes on the whole of* S_2.

Proof. From (5.1.3), by the method of linear transformation of equations (Krein, 1971, §10), we obtain

$$\int_{S_2} G(p, E) f(p) \, \mathrm{d}_p S_2 = \int_{S_2} G(p, E) \Big(\int_{D_1} G(p, Q) v(Q) \, \mathrm{d}Q \Big) \, \mathrm{d}_p S_2, \quad (5.1.4)$$

where v is a new unknown function harmonic in D_1. Interchanging the integration over S_2 with integration over D_1 in (5.1.4) and introducing the notations

$$\Psi(E) = \int_{S_2} G(p, E) f(p) \, \mathrm{d}_p S_2,$$

$$G_{S_2}(E, Q) = \int_{S_2} G(p, E) G(p, Q) \, \mathrm{d}_p S_2,$$

we obtain

$$\Psi(E) = \int_{D_1} G_{S_2}(E, Q) v(Q) \, \mathrm{d}Q. \quad (5.1.5)$$

Now, instead of (5.1.5), let us consider the equation iterated ν times ($\nu \geq [m/2]$):

$$\Psi_\nu(E_\nu) = \int_{D_1} G_{S_2\nu}(E_\nu, Q) v(Q) \, \mathrm{d}Q, \qquad E_\nu \in D_1. \quad (5.1.6)$$

Since $G_{S_2\nu}(E_\nu, Q)$ is expanded into a uniformly converging series in the eigenfunctions $\{u_n\}_1^\infty$ of the Laplace operator for the Dirichlet problem in the domain D_1 (Il'in, 1958), therefore, we obtain from (5.1.6):

$$\Psi_\nu(E_\nu) = \sum_{n=1}^{\infty} C_n u_n(E_\nu). \quad (5.1.7)$$

Here

$$C_n = (\Psi_\nu, u_n)_{D_1} = \sum_{k=1}^{n} \lambda_k^{-\nu} \lambda_{n-k+1}^{-1} C'_{n-k+1} l_{n,n-k+1} \qquad (n = 1, 2, \dots),$$

λ_n are the eigenvalues of the eigenfunctions u_n, and

$$C'_{n-k+1} = \int_{D_1} u_{n-k+1}(Q) v(Q) \, \mathrm{d}Q,$$
$$l_{n,n-k+1} = \int_{S_2} u_n(p) u_{n-k+1}(p) \, \mathrm{d}_p S_2.$$

Evidently, $C_n = 0$ for $n = 1, 2, \dots$, because $\Psi_\nu(E_\nu) = 0$ for $f(p) = 0$. This holds for $\varphi_\alpha = 0$ $(\alpha = 1, 2)$. In this case (see Kostrikin, 1977, p. 112), we have

$$C'_n = \int_{D_1} u_n(Q) v(Q) \, \mathrm{d}Q = 0. \tag{5.1.8}$$

Since the system $\{u_n\}_1^\infty$ is complete in $L_2(D_1)$, therefore, (5.1.8) implies $v(Q) = 0$ for $Q \in D_1$.

Finally, $u(p) = 0$ for $p \in D_1$ since

$$u(p) = -\int_{D_1} G(p, Q) v(Q) \, \mathrm{d}Q + \int_{S_1} \frac{\partial G}{\partial n_Q} \varphi_1(Q) \, \mathrm{d}_Q S_1.$$

□

Theorem 5.1.2. *The solution of the conjugate problem is also unique in the class $C^2(\overline{D_1}) \cap C^4(D_1)$ if the conditions of Theorem 5.1.1 hold.*

Proof. Really, the integral Fredholm equation of first kind in the class of functions with limited variation

$$\int_{S_2} \nu(p) K(p, q) \, \mathrm{d}_p S_2 = \Psi(q), \quad q \in S_1, \tag{5.1.9}$$

where

$$K(p, q) = \int_{D_1} G(p, Q) \frac{\partial G(q, Q)}{\partial n_Q} \, \mathrm{d}Q,$$

is the equation adjoint to equation (5.1.3).

Consider the function

$$v(q) = \int_{S_2} \nu(p) K_1(p, q) \, \mathrm{d}_p S_2, \tag{5.1.10}$$

where

$$K_1(p,q) = \int_{D_1} G(p,Q)G(q,Q)\,\mathrm{d}Q.$$

We see that the problem

$$\Delta^2 v(q) = 0, \qquad q \in (D_1 \setminus \overline{D_2}) \cup D_2, \tag{5.1.11}$$

$$v(q) = \frac{\partial}{\partial n_q} v(q) = \Delta v(q) = 0,$$

$$\frac{\partial}{\partial n_q} \Delta v(q) = \int_{S_2} v(p) \frac{\partial G}{\partial n_q}\,\mathrm{d}_p S_2 \tag{5.1.12}$$

is equivalent to the integral equation (5.1.9).

Hence, $\nu(p) = 0$ if the conditions of Theorem 5.1.1 hold, and $v(q) = 0$, $q \in D_1$, since the solution of the problem (5.1.11), (5.1.12) is unique. □

Note that

$$\int_{S_2} u_1(p) u_n(p)\,\mathrm{d}_p S_2 = 0$$

for at least one n, for example, when $u_n(p)$ vanishes on the whole of S_2 (Vladimirov, 1984). In this case C_n' may be distinct from zero, and the solution of the problem (5.1.1), (5.1.2) as well as the solution of the problem (5.1.11), (5.1.12) will be not unique.

5.2. APPENDIX B

Here we apply the method of linear change of unknowns (Krein, 1971, §15) to establish the uniqueness of solutions of the first internal boundary value problem and of the problem conjugate to it for the biharmonic equation in the class $C^2(\overline{D_1}) \cap C^4(D_1)$ of m-dimensional functions ($m \geq 2$).

1. Let D_α ($\alpha = 1, 2$) be finite domains of the Euclidean space $\mathbb{R}^m$ ($m \geq 2$) bounded by the Lyapunov surfaces S_α, $\alpha = 1, 2$, respectively, and let $D_1 \supset \overline{D_2}$.

Two-surface problem. Find a function $u(p)$, $p = (x_1, \dots, x_m) \in D_1$, satisfying the conditions

$$\Delta^2 u(p) = 0, \qquad p \in D_1, \tag{5.2.1}$$

$$u(p) = \varphi_\alpha(p), \qquad p \in S_\alpha \qquad (\alpha = 1, 2), \tag{5.2.2}$$

where Δ is the Laplace operator and φ_α are given functions.

One can see that the problem (5.2.1), (5.2.2) contains the following integral Fredholm equation of first kind

$$\int_{S_1} K(p,q)\Psi(q)\,\mathrm{d}_q S_1 = f(p), \qquad p \in S_2, \tag{5.2.3}$$

where

$$K(p,q) = \sum_{n=1}^{\infty} \frac{u_n(p)}{\lambda_n^2} \frac{\partial u_n(q)}{\partial n_q}, \qquad f(p) = \varphi_2(p) - \int_{S_1} \frac{\partial G(p,Q)}{\partial n_Q} \varphi_1(Q)\,\mathrm{d}_Q S_1.$$

Here u_n are the eigenfunctions of the Laplace operator for the Dirichlet problem corresponding to the eigenvalues λ_n (Il'in, 1958), and G is the Green function. Interchanging in (5.2.3) the integration and summation we obtain

$$\sum_{n=1}^{\infty} \frac{u_n(p)}{\lambda_n^2} C_n = f(p), \qquad p \in S_2, \tag{5.2.4}$$

where

$$C_n = \int_{S_1} \frac{\partial u_n(q)}{\partial n_q} \Psi(q)\,\mathrm{d}_q S_1, \qquad u_n(p) \neq 0.$$

2. Uniqueness of C_n. Following the method of linear change of variables (Krein, 1971), we set

$$C_n = \lambda_n d_n. \tag{5.2.5}$$

Then (5.2.4) yields

$$\sum_{n=1}^{\infty} \frac{u_n(p)}{\lambda_n} d_n = f(p), \qquad p \in S_2. \tag{5.2.6}$$

A similar relation can be obtained from the equation

$$\int_{S_1} \frac{\partial G}{\partial n_q} Z(q)\,\mathrm{d}_q S_1 = f(p), \qquad p \in S_2, \tag{5.2.7}$$

where

$$\frac{\partial}{\partial n_q} G(p,q) = \sum_{n=1}^{\infty} \frac{u_n(p)}{\lambda_n} \frac{\partial u_n(q)}{\partial n_q}. \tag{5.2.8}$$

Comparing (5.2.6) and (5.2.7) we find

$$d_n = \int_{S_1} \frac{\partial u_n(q)}{\partial n_q} Z(q)\, \mathrm{d}_q S_1. \tag{5.2.9}$$

The uniqueness of $Z(q)$ for $q \in S_1$ follows from (5.2.7). Thus, (5.2.5) implies, since $\lambda_n > 0$, the uniqueness of C_n for each n and, consequently, the uniqueness of $\Psi(q)$ in (5.2.3).

3. The theorem below is a corollary of the results of items 1 and 2.

Theorem 5.2.1. *The solution of the problem* (5.2.1), (5.2.2) *is unique in the class* $C^2(\overline{D_1}) \cap C^4(D_1)$ *if* $u_n(p) \neq 0$ *for all* n *and* $p \in S_2$.

4. We consider the equation adjoint to (5.2.3)

$$\int_{S_2} K(p,q) g(p)\, \mathrm{d}_p S_2 = l(q), \qquad q \in S_1, \tag{5.2.3*}$$

where $g(p)$ is a function with bounded variation in S_2. From (5.2.3*), using the expansion of $K(p,q)$ in (5.2.3) and taking $g(p) \in L(S_2)$, we obtain

$$\sum_{n=1}^{\infty} \frac{1}{\lambda_n^2} \frac{\partial u_n(q)}{\partial n_q} g_n = l(q), \qquad q \in S_1, \tag{5.2.4*}$$

where $g_n = \int_{S_2} u_n(p) g(p)\, \mathrm{d}_p S_2$.

5. Uniqueness of g_n. Let (Krein, 1971)

$$g_n = \lambda_n l_n. \tag{5.2.5*}$$

Substituting (5.2.5*) into (5.2.4*), we find

$$\sum_{n=1}^{\infty} \frac{1}{\lambda_n} \frac{\partial u_n(q)}{\partial n_q} l_n = l(q), \tag{5.2.6*}$$

where $l_n = \int_{S_2} u_n(p) Z(p)\, \mathrm{d}_p S_2$ and $Z(p)$ is an integrable function on S_2.

We see that we may derive (5.2.6*) from the equation

$$\int_{S_2} \frac{\partial G}{\partial n_q} Z(p)\, \mathrm{d}_p S_2 = l(q), \qquad q \in S_1. \tag{5.2.10}$$

Let us consider the following potential of the Green simple layer

$$W(q) = \int_{S_2} G(p,q)Z(p)\,\mathrm{d}_pS_2. \tag{5.2.11}$$

It easy to see that

$$\begin{gathered} \Delta W = 0 \quad \text{in} \quad (D_1 \setminus \overline{D_2}) \cup D_2, \\ W(q) = 0, \quad \frac{\partial W(q)}{\partial n_q} = l(q) \quad \text{in} \quad S_1, \end{gathered} \tag{5.2.12}$$

since the solution to the problem (5.2.12) is unique in D_1 because (5.2.11) is continuous when W crosses S_2. This implies the uniqueness of $Z(p)$ in S_2 for all Lebesgue points (Gyunter, 1953) from the formula of jumps of normal derivatives of the potential (5.2.11) when we cross S_2.

Now, uniqueness of $g(p)$ follows from uniqueness of l_n from (5.2.5*) and (5.2.6*).

6. The next theorem is a consequence of the results of items 4 and 5.

Theorem 5.2.2. *The solution of the problem conjugate to* (5.2.1), (5.2.2) *is also unique if the conditions of Theorem 5.2.1 hold.*

Note that, if $u_n(p) = 0$, $p \in S_2$, for a certain n, then the solution of the problem (5.2.1), (5.2.2) and the problem conjugate to it may be not unique.

Conclusion. If the surface S_2 distinct from a sphere is not a zero of any eigenfunction of the Laplace operator, then the first internal boundary value problem for the biharmonic equation is completely solvable.

5.3. APPENDIX C

Here we consider the first internal boundary value problem in a spatial star-like domain.

Let D_α ($\alpha = 1, 2$) be finite domains in the Euclidean space $\mathbb{R}^m$ ($m \geq 2$). We assume that D_α are star-like relative to the origin and are bounded by the Lyapunov surfaces S_α ($\alpha = 1, 2$), respectively. The equations for these surfaces are $r = R_\alpha(\theta)$, $\alpha = 1, 2$; $R_1(\theta) \geq R_2(\theta)$.

Problem. We need to find a function $u(p)$, $p = (r, \theta) \in D_1$, satisfying the conditions

$$\Delta^2 u(p) = 0, \qquad p \in D_1, \tag{5.3.1}$$

$$u(p) = \varphi_\alpha(p), \qquad p \in S_\alpha \qquad (\alpha = 1, 2). \tag{5.3.2}$$

Here Δ is the Laplace operator, and φ_α $(\alpha = 1, 2)$ are continuous functions given on S_α $(\alpha = 1, 2)$, respectively.

Theorem 5.3.1. *If one of the surfaces S_α $(\alpha = 1, 2)$ is a sphere, then the first internal boundary value problem is uniquely completely solvable in the class $C^2(\overline{D_1}) \cap C^4(D_1)$.*

Proof. We use the following representation for a biharmonic function u in a star-like domain D_1 (Tikhonov and Samarskii, 1990):

$$u(r, \theta) = u_1(r, \theta) + r^2 u_2(r, \theta), \tag{5.3.3}$$

where

$$u_\alpha(p) = \int_{S_1} \mu_\alpha(q) \frac{\partial}{\partial n_q} G(p, q) \, \mathrm{d}_q S_1, \tag{5.3.4}$$

and G is the Green function of the Dirichlet problem for the Laplace equation.

Taking into account condition (5.3.2) we pass from the problem (5.3.1), (5.3.2) to the equivalent system of integral Fredholm equations of third kind

$$\mu_1(p) + R_1^2(p)\mu_2(p) = \varphi_1(p), \qquad p \in S_1, \tag{5.3.5}$$

$$\int_{S_1} \mu_1(q) \frac{\partial}{\partial n_q} G(p, q) \, \mathrm{d}_q S_1 + R_1^2(p) \int_{S_1} \mu_2(q) \frac{\partial}{\partial n_q} G(p, q) \, \mathrm{d}_q S_1 = \varphi_2(p),$$
$$p \in S_2. \tag{5.3.6}$$

Really, assume that the functions $\mu_\alpha(q) \in C(S_1)$ solve the system (5.3.5), (5.3.6). Then solving the Dirichlet problems relative to $\mu_\alpha(q)$, $\alpha = 1, 2$, we find (5.3.4). Substituting (5.3.4) into (5.3.3) we find the solution of the problem (5.3.1), (5.3.2). Conversely, if a function $u(p) \in C^2(\overline{D_1}) \cap C^4(D_1)$ solves the problem (5.3.1), (5.3.2), then it can be represented in the form (5.3.3).

Taking into account (5.3.3) and the fact that D_1 is a star-like domain, we conclude that

$$u_1(R_1(\theta), \theta) = u(R_1(\theta), \theta) - R_1^2(\theta) u_2(R_1(\theta), \theta) = \mu_1(q) \in C(S_1),$$
$$u_2(R_1(\theta), \theta) = \frac{1}{4} R_1^{-m/2} \int_0^{R_1(\theta)} \tau^{(m/2)-1} \Delta u \, \mathrm{d}\tau = \mu_2(q) \in C(S_1)$$

solve the system (5.3.5), (5.3.6).

Let S_2 be a sphere. To prove the uniqueness of $\mu_\alpha(q)$, $\alpha = 1,2$, we consider the homogeneous system

$$\mu_1(p) + R_1^2(p)\mu_2(p) = 0, \qquad p \in S_1,$$
$$\int_{S_1} \mu_1(q)\frac{\partial}{\partial n_q}G(p,q)\,\mathrm{d}_qS_1 + R_2^2\int_{S_1} \mu_2(q)\frac{\partial}{\partial n_q}G(p,q)\,\mathrm{d}_qS_1 = 0, \qquad p \in S_2, \tag{5.3.7}$$

corresponding to the system (5.3.5), (5.3.6). Eliminating μ_1 from system (5.3.7) we obtain the following homogeneous integral Fredholm equation of first kind for the function μ_2:

$$\int_{S_1} (R_1^2(q) - R_2^2)\mu_2(q)\frac{\partial G}{\partial n_q}\,\mathrm{d}_qS_1 = 0, \qquad p \in S_2. \tag{5.3.8}$$

Hence, $\mu_2 = 0$. Then from the first equation of system (5.3.7) we get $\mu_1 = 0$.

Now we prove that the solution of the system conjugate to (5.3.7)

$$\nu_1(q) + \int_{S_2} \nu_2(p)\frac{\partial}{\partial n_q}G(p,q)\,\mathrm{d}_pS_2 = 0, \qquad q \in S_1,$$
$$\nu_1(q)R_1^2(q) + R_2^2\int_{S_2} \nu_2(p)\frac{\partial}{\partial n_q}G(p,q)\,\mathrm{d}_pS_2 = 0, \qquad q \in S_1, \tag{5.3.9}$$

is also trivial in the class of functions $\nu_\alpha(q)$, $q \in S_\alpha$, with bounded variation.

From (5.3.9) for $\nu_2 \in L(S_2)$ we obtain

$$(R_1^2(q) - R_2^2)\int_{S_2} \nu_2(p)\frac{\partial}{\partial n_q}G(p,q)\,\mathrm{d}_pS_2 = 0, \qquad q \in S_1 \tag{5.3.10}$$

Since $R_1(q) > R_2$, therefore, (5.3.10) implies

$$\int_{S_2} \nu_2(p)\frac{\partial}{\partial n_q}G(p,q)\,\mathrm{d}_pS_2 = 0, \qquad q \in S_1. \tag{5.3.11}$$

Consider the function

$$\nu_2(q) = \int_{S_2} \nu_2(p)G(p,q)\,\mathrm{d}_pS_2 = 0, \qquad q \in D_1. \tag{5.3.12}$$

Evidently,

$$\Delta^2\nu_2(q) = 0, \qquad q \in D_1 \setminus \overline{D_2},$$

$$\nu_2(q) = 0, \qquad \frac{\partial}{\partial n_q}\nu_2(q) = 0, \qquad q \in S_1.$$

Then, since the Cauchy problem for the Laplace equation has a unique solution in $D_1 \setminus \overline{D_2}$, we conclude that $\nu_2(q) = 0$, $q \in D_1 \setminus \overline{D_2}$.

The properties of the limit passage of $\nu_2(q)$ across S_2 (Gyunter, 1953, p. 134) imply that

$$\nu_2(q) = 0, \qquad q \in D_2.$$

Then the well-known formula for jumps of normal derivatives of the potential of the Green simple layer (5.3.12) implies that $\nu_2(q) = 0$, $q \in S_2$, in $L(S_2)$ at all the Lebesgue points q (Gyunter, 1953, p. 132) of the integrable function $\nu_2(q)$.

Finally, (5.3.9) implies that $\nu_1(q) = 0$, $q \in S_1$. □

Bibliography

Bochner, S. (1959). *Lectures on Fourier Integrals.* Annals of Mathematics Studies, No. 42. Princeton University Press, Princeton, New Jersey.

Gyunter, N. M. (1953). *Potential Theory and Its Application to Basic Problems of Mathematical Physics.* Gosteorizdat, Moscow (in Russian).

Il'in, V. A. (1958). On uniform convergence of expansions by eigenfunctions in the whole closed domain. *Matem. Sbornik,* 45 (2), 195–232 (in Russian).

Ivanov, V. K. (1965). The Cauchy problem for the Laplace equation in an infinite strip. *Differ. Uravn.,* 1 (1), 131–135 (in Russian).

Koshlyakov, N. S., Gliner, E. B., and Smirnov, M. M. (1970). *Equations in Partial Derivatives of Mathematical Physics.* Vyssh. Shkola, Moscow (in Russian).

Kostrikin, A. I. (1977). *Introduction into Algebra.* Nauka, Moscow (in Russian).

Krein, S. G. (1971). *Linear Equations in a Banach Space.* Nauka, Moscow (in Russian).

Lavrent'ev, M. A., and Shabat, B. V. (1966). *Methods of the Theory of Functions of Complex Variable.* Nauka, Moscow (in Russian).

Lavrent'ev, M. M. (1962). *On Some Ill-Posed Problems of Mathematical Physics.* Siberian Branch of the USSR Acad. Sci., Novosibirsk (in Russian).

Lavrent'ev, M. M. (1973). *Conditionally Well-Posed Problems for Differential Equations.* Novosibirsk State University, Novosibirsk (in Russian).

Lavrent'ev, M. M. (1981). *Ill-Posed Problems for Differential Equations.* Novosibirsk State University, Novosibirsk (in Russian).

Leontiev, A. F. (1976). *Exponential Series.* Nauka, Moscow (in Russian).

Shishatskii, S. P. (1979). On zeros of periodic harmonic functions. In: *Conditionally Well-Posed Mathematical Problems and Problems of Geophysics.* Computing Center of Siberian Branch of the USSR Acad. Sci., Novosibirsk, 126–133 (in Russian).

Sobolev, S. L. (1950). *Some Applications of Functional Analysis in Mathematical Physics.* Leningrad State University, Leningrad (in Russian).

Tikhonov, A. N., and Arsenin, V. Ya. (1977). *Solutions of Ill-Posed Problems.* V. H. Winston & Sons, Washington, D.C.; John Wiley & Sons, New York.

Tikhonov, A. N., and Samarskii, A. A. (1990). *Equations of Mathematical Physics.* Dover Publications, Inc., New York.

Tricomi, F. (1957). *Integral Equations.* Pure and Applied Mathematics, Vol. V. Interscience Publishers, New York – London.

Vladimirov, V. S. (1984). *Equations of Mathematical Physics.* Mir, Moscow.

Zigmund, A. (1959). *Trigonometric Series. Vol. I.* Cambridge at the University Press.

Bei Fragen zur Produktsicherheit wenden Sie sich bitte an:
If you have any questions regarding product safety,
please contact:

Walter de Gruyter GmbH
Genthiner Straße 13
10785 Berlin
productsafety@degruyterbrill.com